Yefeng Zheng
David Doermann
Huiping Li

Handwritten Document Image Processing

Yefeng Zheng
David Doermann
Huiping Li

Handwritten Document Image Processing

Identification, Matching, and Indexing of Handwriting in Noisy Document Images

VDM Verlag Dr. Müller

Impressum/Imprint (nur für Deutschland/ only for Germany)

Bibliografische Information der Deutschen Nationalbibliothek: Die Deutsche Nationalbibliothek verzeichnet diese Publikation in der Deutschen Nationalbibliografie; detaillierte bibliografische Daten sind im Internet über http://dnb.d-nb.de abrufbar.
Alle in diesem Buch genannten Marken und Produktnamen unterliegen warenzeichen-, marken- oder patentrechtlichem Schutz bzw. sind Warenzeichen oder eingetragene Warenzeichen der jeweiligen Inhaber. Die Wiedergabe von Marken, Produktnamen, Gebrauchsnamen, Handelsnamen, Warenbezeichnungen u.s.w. in diesem Werk berechtigt auch ohne besondere Kennzeichnung nicht zu der Annahme, dass solche Namen im Sinne der Warenzeichen- und Markenschutzgesetzgebung als frei zu betrachten wären und daher von jedermann benutzt werden dürften.

Coverbild: www.purestockx.com

Verlag: VDM Verlag Dr. Müller Aktiengesellschaft & Co. KG
Dudweiler Landstr. 125 a, 66123 Saarbrücken, Deutschland
Telefon +49 681 9100-698, Telefax +49 681 9100-988, Email: info@vdm-verlag.de
Zugl.: College Park, University of Maryland, Diss., 2006

Herstellung in Deutschland:
Schaltungsdienst Lange o.H.G., Zehrensdorfer Str. 11, D-12277 Berlin
Books on Demand GmbH, Gutenbergring 53, D-22848 Norderstedt
Reha GmbH, Dudweiler Landstr. 99, D- 66123 Saarbrücken
ISBN: 978-3-639-09192-2

Imprint (only for USA, GB)

Bibliographic information published by the Deutsche Nationalbibliothek: The Deutsche Nationalbibliothek lists this publication in the Deutsche Nationalbibliografie; detailed bibliographic data are available in the Internet at http://dnb.d-nb.de.
Any brand names and product names mentioned in this book are subject to trademark, brand or patent protection and are trademarks or registered trademarks of their respective holders. The use of brand names, product names, common names, trade names, product descriptions etc. even without
a particular marking in this works is in no way to be construed to mean that such names may be regarded as unrestricted in respect of trademark and brand protection legislation and could thus be used by anyone.

Cover image: www.purestockx.com

Publisher:
VDM Verlag Dr. Müller Aktiengesellschaft & Co. KG
Dudweiler Landstr. 125 a, 66123 Saarbrücken, Germany
Phone +49 681 9100-698, Fax +49 681 9100-988, Email: info@vdm-verlag.de

Produced in USA and UK by:
Lightning Source Inc., 1246 Heil Quaker Blvd., La Vergne, TN 37086, USA
Lightning Source UK Ltd., Chapter House, Pitfield, Kiln Farm, Milton Keynes, MK11 3LW, GB
BookSurge, 7290 B. Investment Drive, North Charleston, SC 29418, USA
ISBN: 978-3-639-09192-2

TABLE OF CONTENTS

LIST OF TABLES

LIST OF FIGURES

Chapter 1

Introduction

Handwriting was developed as a means to expand human memory and to facilitate communication. It has changed tremendously over time when new writing tools are invented. The latest change is, that with the acceptance of new technologies such as personal digital assistants (PDAs) and cellular phones, handwriting can be collected on-line without losing temporal information, which opens a great opportunity for the analysis of handwriting, such as handwriting recognition and signature verification. New technologies also challenge the persistence of handwriting. For example, the oldest books are hand-copied. However, the printing press and typewriters opened up the world of formatted documents and made scriptoria obsolete. Almost all books in the past several centuries have been machine printed. Recently, computer and communication technologies such as word processors, fax machines, and e-mails provide new ways to expand human memory as well as facilitate communication. In this perspective, one may ask: Will handwriting be threatened with extinction?

All these inventions have led to the fine-tuning and reinterpreting of handwriting. With the increase of literacy, more and more people learn to read and write. As a general rule, as the length of the handwritten message decreases, the number of people using handwriting increases [4]. Widespread acceptance of digital computers seemingly challenges the future of handwriting. However in numerous situations, pen and paper provides more convenience than a keyboard. For example, most students still do not type lecture notes on a notebook computer. They record language, equations, and graphs with a pen. Many people still prefer to keep a hard copy of documents, even when electronic versions are available. They make annotations on a document when they are reading. Also, handwriting is demanded by law such as signatures on legal documents. This brings a new challenge to process such documents where handwritten annotations mix with machine printed contents. Since the segmentation and recognition techniques for machine printed text and handwriting are very different, different contents should be identified before further processing.

Documents are the result of a set of physical processes and conditions, and the resulting document can be viewed as consisting of layers, such as handwriting, machine printed text, background pattern, figures, tables, and/or noise. Fig. 1.1a shows how several layers combine to generate a document. Not all layers are present in a single document. The dashed arrows in the figure mean that the corresponding components are missed in this example. Document analysis reverses these processes to segment a document into layers with different physical and semantic properties. This procedure is shown in Fig. 1.1b. Since the 1960s, much research on document processing has been done based on optical character recognition (OCR). A more general study of document analysis, such as page (or zone) segmentation, zone classification, and table detection, began in early 1980s. After more than two decades of research, automatic machine printed text segmentation and recognition for clean documents can be viewed as a solved problem with commercial products on the market. However, much work needs to be done for handwriting, such as separating

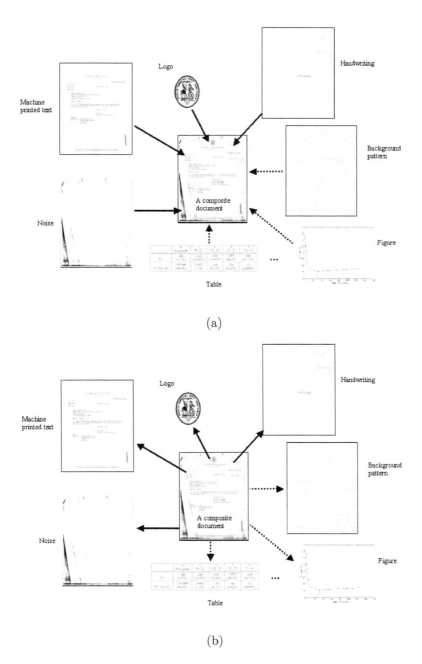

(a)

(b)

Figure 1.1: (a) A document generation model. (b) A document image analysis model.

handwriting from machine printed text, segmentation and recognition of handwriting.

The study of handwriting covers a broad field, dealing with numerous aspects of this complex task. It involves research concepts from several disciplines: experimental psychology [5], computer science [6], education [7], and forensic document examination [8]. For computer processing of handwriting there are several types of analysis, recognition, and interpretation associated with it. Handwriting recognition transforms the spatial form of graphical marks into symbolic representation. Signature identification determines the author of a sample from a set of individuals. Signature verification determines whether the signature belongs to a given person.

This work presents our approach to identifying the handwriting layer in a document image from other layers such as background patterns, noise, and machine printed text. After handwriting identification, we propose an approach to handwriting matching that can be applied for handwriting synthesis and retrieval.

1.1 Rule Line Detection

Many handwritten documents come together with a special background pattern: rule lines. These rule lines are printed on the paper to guide writing. After digitization they will, however, touch text and cause problems for further document analysis such as segmentation and recognition. These lines must be detected and removed before the text is fed to an optical character recognition (OCR) engine. These rule lines may appear severely broken since they are thin and printed with a light color. Many line detection algorithms have been proposed in the literature [9, 10, 11, 12, 13, 1]. They work well on relatively clean documents with solid or mildly broken lines, but performance will significantly deteriorates if lines are severely broken because of low image quality or if they mix, touch, and overlap with text. It is difficult, if not impossible, to reliably detect these lines individually. Another challenge involves character strokes, which may lie on the same line, causing a high false alarm rate. If the false alarm lines are removed, character strokes may be removed erroneously. A line detection algorithm with both a high accuracy and a low false alarm rate should be developed for severely broken rule lines.

To handle these problems, the context is often required to refine the initial detection. For example, in form processing most form cells are rectangular. In known form processing the number of lines and the gaps among these lines can be used as a priori knowledge and stored as references in form templates. These ideas have been presented in previous work to improve detection accuracy and reduce false alarms [14, 1, 15, 16]. But the usage of a priori knowledge in the above applications is ad hoc and lacks a systematic representation.

In this work, we present a rule line detection algorithm based on hidden Markov model (HMM) decoding. After skew estimation and correction, we perform a horizontal projection. An HMM model is used to model the projection profile, and the positions of all rule lines are detected simultaneously after HMM decoding with the Viterbi algorithm. Experiments on a real data set show that our algorithm achieves both a high detection accuracy and a low false alarm rate. After detection, lines are removed by line width thresholding.

1.2 Handwriting Identification in Noisy Documents

Handwriting often mixes with machine printed text. Handwriting in a machine printed document often indicates corrections, additions, or other supplemental information that should be treated differently from the main content. The segmentation and recognition techniques required for machine printed and handwritten text differ significantly. Therefore, identification of handwriting from machine printed text is crucial for the following document image analysis.

The data set we are processing is noisy, which makes the problem more challenging. Large (e.g., marginal black strips) and small noise components (e.g., pepper-and-salt noise) can be removed reliably with some simple rules [17, 18]. It is, however, hard to discriminate noise from compatible sized text. In this work, we treat noise as a distinguished class. We first segment the document at a suitable level, and each segmented block is classified into machine printed text, handwriting, or noise.

Some work has been done on handwriting/machine printed text identification. The classification is typically performed at the text line [19, 20, 21, 22], word [23], or character level [24, 25]. Special consideration must be given to the size of the region being segmented before we can perform any classification. The smallest unit used for classification is called the *pattern unit*. If the unit is too small, the information contained in it may not be sufficient for classification; if it is too large, however, different types of components may be mixed in the same region due to segmentation errors. In previous work, we conducted a performance evaluation for the accuracy of machine printed text/handwriting distinguish at the character, word, and zone levels, and showed that a reliable classification can be achieved at the word level [25]. Several features, such as Gabor filter features, run-length histogram features, crossing counts histogram features, and texture features, are extracted to identify each segmented block into machine printed text, handwriting, or noise.

Several classifiers, such as the Fisher linear discriminant classifier, the k-nearest neighbor (k-NN) classifier, and the support vector machine (SVM) classifier, are tested in our comparison experiments. They have similar performance with reasonable accuracy. If the machine printed text block is too small (such as words with less than 3 characters), it is likely to be classified as noise. Some noise blocks are classified as handwriting due to the overlapping in the feature space of these two classes. Machine printed text, handwriting, and noise exhibit different patterns of geometric relationships. For example, printed words often form horizontal (or vertical) text lines, while noise blocks tend to overlap each other. Markov random field (MRF) is used to model such geometric relationships to refine the classification results. Experiments show MRF is effective in modeling the geometric dependency of neighboring components, about half of the mis-classifications are corrected after post-processing. After identification, noise can be removed from the document, which enhances a degraded document. Machine printed text can be sent for zone segmentation or recognition with any off-the-shelf OCR package. Identified handwriting can be sent for further analysis, such as recognition, retrieval, signature verification or identification.

7

1.3 Handwriting Matching, Synthesis, and Retrieval

The identified handwriting may be sent for further analysis. In this work, we propose a novel handwriting matching technique and apply it for handwriting synthesis and retrieval. We study handwriting matching in a broader context of nonrigid shape matching using a set of points uniformly sampled from the handwriting skeleton. For nonrigid shapes, most neighboring points cannot move independently under deformation due to physical constraints. Therefore, though the absolute distance between two points may change significantly, the neighborhood of a point is well preserved in general. Based on this observation, we formulate point matching as an optimization problem to preserve local neighborhood structures during matching. Our formulation has a simple graph matching interpretation, where each point is a node in the graph, and two nodes are connected by an edge if they are neighbors. The optimal match between two graphs is the one that maximizes the number of matched edges (i.e., the number of neighborhood relations). The shape context distance is used to initialize the graph matching, followed by relaxation labeling for refinement. Experiments demonstrate the effectiveness of our approach: it outperforms the shape context [26] and TPS-RPM [2] algorithms under nonrigid deformation and noise on a public data set.

The performance of a statistical pattern recognition system depends heavily on the size and quality of the training set. Although it is easy to prepare samples of machine printed text, doing so is expensive for handwriting. Synthesized data can be used as a supplement. The key problem of handwriting synthesis is generating sam-

8

ples that look natural. Otherwise, arbitrarily synthesized samples cannot improve (if not deteriorate) the performance of the system trained on them. Although handwriting samples vary greatly in respect to size, rotation, and stroke width, shape is generally used to categorize them into different classes. Since nonrigid deformation of handwriting is large, we argue that a synthesis algorithm should learn the shape deformation characteristics from real handwriting samples. It is reasonable to assume that the shape space of handwriting with the same content (e.g., the handwriting samples of the letter 'a') is continuous. For characters with several different writing glyphs, such as number '7,' we may need to do clustering analysis to segment the shape space into multiple continuous sub-space. Given two handwriting samples close in the shape space, an interpolation between them is likely to lie inside the shape space too (this is guaranteed if the shape space is convex). That means, given two real similar handwriting samples, it is reasonable to assume some person may write with a shape between them (i.e., with similar but less degree of deformation). In this work, we propose an example-based handwriting synthesis approach using two training samples. We use our handwriting matching algorithm to establish the correspondence between two handwriting samples. After handwriting matching, we warp one sample toward the other using the thin plate spline (TPS) deformation model. By adjusting the regularization parameter of the TPS deformation model, we can control the amount of nonrigid deformation in synthesis.

Another application of handwriting matching is handwriting retrieval. Recently, shape context [26] was proposed as an effective tool for shape recognition and retrieval. In this approach, the point correspondence is estimated, and similarity measures are defined based on the matching result for shape retrieval. By replacing the original shape matching method with our more robust approach, we achieve moderate improvement. To further improve the accuracy, we propose new similarity measures, such as a measure based on the affine transformation, registration residual errors, and outlier ratio estimated by the matching algorithm. Using more similarity measures will significantly improve the retrieval accuracy. A more effective way to improve accuracy is to use multiple query samples. We propose a simple but effective way to combine the retrieval results using multiple query instances.

1.4   Organization of the Book

This book is organized as following: Rule line detection and removal is described in detail in the next chapter. Our model-based line detection algorithm is not limited to handwriting document analysis. In Chapter 3, we apply it for known form processing. In Chapter 4, we present our approach to identify handwriting and machine printed text in noisy document images. Our handwriting matching approach is described in Chapter 5. We discuss the application of handwriting matching to handwriting synthesis and handwriting retrieval in Chapters 6 and 7, respectively. This book concludes with a brief summary of our contributions, and some discussions of the remaining problems.

Chapter 2

Rule Line Detection

2.1 Introduction

Many handwritten documents come with a special background pattern, rule lines, as shown in Fig. 2.1a. It is important that these lines are detected and removed before the text goes to an optical character recognition (OCR) engine. In this chapter, we focus on rule line detection.

Many line detection algorithms have been proposed in the literature [9, 10, 11, 12, 13, 1]. They work well on relatively clean documents with solid or mildly broken lines, but the performance will deteriorate significantly if lines are severely broken due to the low image quality, or if they mix, touch, and overlap with text. Fig. 2.1b shows the line detection results for a rule-lined document using the directional singly-connected chain (DSCC) method [1]. We can see only a few lines are partially detected due to severe brokenness. It is very difficult, if not impossible, to reliably detect these lines individually.

In this work, we propose a model-based method which incorporates context to detect parallel lines optimally and systematically. Under the model, lines are detected by a hidden Markov model (HMM) decoding process, which can determine the positions of all lines simultaneously. Rather than detecting lines directly on original images [10, 13, 1], we use a DSCC-based scheme to filter text as a preprocessing

11

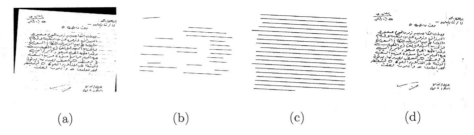

| (a) | (b) | (c) | (d) |

Figure 2.1: Rule line detection and removal. (a) A rule-lined document; (b) Line detection results using the DSCC method [1]; (c) Line detection results using our model based approach; (d) Cleaned image after line removal (the black marginal strips are removed too).

step so the interference with text can be minimized. We then use a coarse-to-fine approach to estimate the skew angle of the document. After deskewing the document, we perform horizontal vertical projection. Rather than treating the peaks in the projection profile as the positions [10, 13], we model the projection profile with an HMM model so the context among these lines can be incorporated. The Viterbi algorithm is then used to search the optimal positions of these lines simultaneously from the projection profile. The experimental results show our method is robust. It can detect lines with a high accuracy and a low false alarm rate in degraded documents. Fig. 2.1c shows the line detection result using the proposed model-based approach. Compared to Fig. 2.1b, our detection result is much better. Our model-based parallel line detection algorithm is flexible; therefore, it can be easily adapted for different applications. We will demonstrate it on two applications: rule line detection and known form processing. In this chapter we focus on rule line detection, and the application on known form processing will be discussed in the next chapter.

After line detection, we would like to remove the detected lines without deteriorating the text. Many line removal algorithms have been developed in the literature, and can be classified into two categories. One kind of approach tries to remove lines completely, then uses local property of overlapping areas, such as stroke direction and connection, to restore the missing parts of strokes [27, 11, 28, 29]. One problem of these approaches is that, after line removal, a large quantity of useful information is lost, making stroke recovery difficult. The other kind of approach analyzes character-line overlapping areas, then removes pixels only belonging to lines, while preserving those belonging to characters [30, 31, 32, 33]. During my graduate work at Tsinghua University in China, I proposed a line width thresholding based approach [34, 35]. The line width is preserved well if no character-line touching happens, but increases noticeably in overlapping areas. This property is used for line removal. We first decompose a line to an array of run-lengths of black pixels, predicated to the running direction of the line. If a run-length is shorter than a threshold, it is removed; otherwise, it is preserved. An adaptive scheme is used to set a small threshold for the area close to text, and a large one for the area far apart from text. Fig. 2.1d shows the line removal result. As we can see, this approach works well for most character-line touching cases. In this chapter, we focus on rule line detection, please refer to our publication [34] for details of our line removal approach.

The remainder of this chapter is organized as follows. In Section 2.2, we briefly review previous work on line detection with emphasis on the usage of prior knowledge. Since text may touch or overlap with lines, removing text before line detection will significantly increase the robustness of our line detection algorithm. In

Section 2.3, we present our text filtering method. Our general model-based parallel

line detection algorithm is described in detail in Section 2.4, which can be tailored

for rule line detection (Section 2.5) or known form processing (discussed in the

next chapter). We demonstrate the robustness of our approach with experiments

in Section 2.6, and the chapter concludes in Section 2.7 with a discussion of future

work.

2.2   Related Work on Line Detection

Line detection is widely used in form detection and interpretation [13, 11, 1], engi-

neering graph interpretation [36], bank check/invoice processing [16, 15], and optical

music recognition (OMR) [37]. Among many algorithms proposed in the literature,

the Hough transform method and its variations are widely used [38, 9]. The Hough

transform method converts the global pattern detection problem in an image space

to a local pattern (ideally a point) detection problem in a transformed parame-

ter space. To detect a straight line, each black pixel (x, y) in an image space is

transformed into a sinusoidal curve in the Hough parameter space

$$\rho = x\cos\theta + y\sin\theta. \tag{2.1}$$

After transformation, collinear points (x_i, y_i) in the image space intersect at a point

(ρ, θ) in the Hough parameter space. Therefore, a peak in the transformed space

provides strong evidence that a corresponding straight line exists in the image. The

Hough transform method can detect dashed and mildly broken lines. However, it

14

is very time consuming. To reduce computational costs, a projection method was proposed [10] to detect form frame lines by limiting the search orientations since only horizontal and/or vertical lines usually exist in form documents. The method deskews the document first, then detects the peaks on the horizontal and vertical projection profiles as lines. It can be viewed as a special case of the Hough transform method by searching θ only around 0° and 90°. The method will fail if the projection of a line does not form a peak on the profile when it mixes with text, the estimated skew angle is not accurate enough, or the lines are too short or severely broken. Chen and Lee [13] proposed the strip projection method to alleviate this problem since lines are more likely to form peaks on the projection profile in a small region. For horizontal line detection, they first divided an image into several vertical strips of equal width, and then performed horizontal projection in each strip. The detected collinear line segments in each strip are linked to form the line.

Thinning is another common method to extract lines. It uses an iterative boundary erosion process to remove outer pixels until only a skeleton of pixel chains remains [39]. It can maintain connectivity, but also tends to create noisy junctions at corners, intersections, and branches. Medial line methods, on the other hand, extract image contours first. The mid-points of two parallel contour lines then form a medial line [40]. The methods may miss pairs of contour lines at branches, requiring post-processing to reduce this distortion [41]. The result of either thinning or medial line method is a chain of pixels, and a line segment can be detected by approximating the pixel chain. The sparse pixel vectorization (SPV) algorithm, proposed by Dori et al. [12], does not use contours to get medial lines. It traces the medial points of consecutive horizontal or vertical pixel runs until constraints are violated. Each continuous trace represents a bar or an arc. SPV often achieves better results than other medial line methods, but the medial point tracking procedure is complicated, and often needs post-processing to refine the results.

Run-lengths are often used as an image component to detect lines. Yu and Jain [11] proposed a data structure, called block adjacency graph (BAG), to represent an image. BAG is defined as $\mathcal{G}(N, E)$, where N is a set of block nodes and E is a set of edges indicating the connection between two nodes. Each node is a block which contains either one or several horizontal run-lengths adjacently connected in the vertical direction and aligned on both left and right sides within a given tolerance. A line is detected by searching a connected sub-graph in the BAG with large aspect ratio. Chhabar et al. [42] presented another run-length-based approach for horizontal line detection. Since the method is composed of four steps: filter, assemble, silhouette, and threshold, they named it the FAST algorithm. The algorithm works directly on run-length encoded images and is very fast. It was later extended to detect lines with any orientation after implementing an efficient rotation operation on run-length coded images [43]. Recently, the directional singly-connected chain (DSCC) method was proposed [1]. A DSCC is a chain of run-lengths which are singly connected. A basic characteristic of a line is that it runs in only one direction. Run-lengths perpendicular to the direction of a line are merged into a DSCC. When a junction is encountered, the merging process stops and a new DSCC generates. Each DSCC represents a line segment, and multiple collinear DSCCs may be merged into a line, based on pre-defined rules. In the above approaches, the group-

16

ing of run-length into line segments is rule-based. A model-based method, using the Kalman filter, was proposed [44]. Assuming that a run-length (perpendicular to the line's running direction) of constant length moves along a straight line, the Kalman filtering technique is used to track the run-length. If the tracking error is larger than a threshold, it is stopped, and a new tracking begins.

In some applications, horizontal and vertical lines always intersect each other. This property can be used to develop an efficient algorithm by detecting intersections of horizontal and vertical lines first. The verification of line segments between intersections complete the algorithm [45, 43].

In most of the above approaches, domain specific knowledge is used implicitly or explicitly. For example, parameters of a line detection algorithm may be tuned to a specified application. In engineering drawing interpretation, knowing the line type (such as solid, dotted, or dashed) helps to develop a robust line detection algorithm [46]. In some forms, most cells are rectangular. This knowledge can be used to improve detection accuracy and reduce false alarms [1]. In [47], Roach and Tatem demonstrated the effectiveness of domain specific knowledge in a highly structured domain: handwritten music score recognition. But the use of the prior knowledge in above applications is ad hoc and lacks systematic representation.

2.3   Preprocessing

Preprocessing has two purposes: first, we deskew the document so the parallel lines are oriented horizontally or vertically; second, we filter text strokes to diminish their intervention in line detection. The skew of a document can be estimated using the text [48], or using the extracted line segments if lines are available on the document [49]. In our approach, we use a coarse-to-fine line based skew estimation method, which is similar to [49]. Since skew estimation is a mature technique in document image analysis, we will not discuss the details in this section. After skew estimation, we can easily rotate the document to correct the skew. In this section, we focus on text filtering, which is one of our contributions. We extract directional singly-connected chains (DSCC) first, then remove DSCCs unlikely to be generated by a line segment because of their shapes.

2.3.1 Definition of DSCC

We define two types of DSCCs: horizontal and vertical, as described in [1]. Take the horizontal DSCC for example. A horizontal DSCC, C_h, consists of a black pixel run-length array $\overline{R_1 R_2 \cdots R_m}$, where R_i is a vertical run-length with one pixel width

$$R_i(x_i, ys_i, ye_i) = \left\{ (x, y) \,\middle|\, \begin{array}{l} p(x, y) = 1, \text{for } x = x_i, y \in [ys_i, ye_i] \\ \text{and } p(x_i, ys_i - 1) = p(x_i, ye_i + 1) = 0 \end{array} \right\}, \quad (2.2)$$

where $p(x, y)$ is the value of pixel (x, y) with 1 representing black pixels, and 0 representing white pixels; x_i, ys_i, and ye_i designate x, starting y, and ending y coordinates of R_i, respectively. Two neighboring run-lengths R_i and R_{i+1} are merged into a DSCC if they are singly connected in the horizontal direction. As shown in Fig. 2.2a, the single connection means that at each side of $R_i(1 < i < m)$, there is one and only one connected run-length. In this example, $\overline{R_1 R_2 \cdots R_7}$, $\overline{R_{11} R_{12} R_{13}}$, R_8, R_9, R_{10}, R_{14} and R_{15} are extracted as DSCCs. The definition of the vertical

18

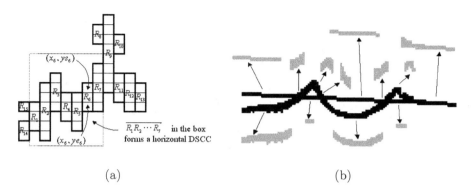

(a)   (b)

Figure 2.2: Definition and extraction of horizontal DSCCs. (a) Illustration of horizontal DSCCs. (b) Extracted DSCCs (represented in gray) where a text stroke crosses a line.

singly-connected chain, C_v, is similar.

The most important property of a line is the single connection along its running direction. An ideal line consists of only one DSCC. A real line often consists of multiple collinear DSCCs. Fig. 2.2b shows an example of extracted DSCCs (represented in gray) of a text stroke crossing a line. We can see the line breaks into several line segments (DSCCs) on the touching area. If the image quality is reasonable, then a line can be detected by merging DSCCs with similar orientation [1]. In our case, we use it to remove text and preserve line segments.

2.3.2 Text Filtering

As shown in Fig. 2.2b, a DSCC can be a text stroke or a line segment. We observed that a line segment often has a smaller variation from the desired orientation and larger aspect ratio. We use an ellipse to model the shape of a DSCC, and calculate

19

the orientation θ, the first and second axes a and b of each DSCC as follows:

$$\mu_{mn} = \sum_x \sum_y (x - \bar{x})^m (y - \bar{y})^n p(x, y) \tag{2.3}$$

$$\theta = 0.5\tan^{-1}\left(\frac{2\mu_{11}}{\mu_{20} - \mu_{02}}\right) \tag{2.4}$$

$$a = \sqrt{\frac{2[\mu_{20} + \mu_{02} + \sqrt{(\mu_{20} - \mu_{02})^2 + 4\mu_{11}^2}]}{\mu_{00}}} \tag{2.5}$$

$$b = \sqrt{\frac{2[\mu_{20} + \mu_{02} - \sqrt{(\mu_{20} - \mu_{02})^2 + 4\mu_{11}^2}]}{\mu_{00}}} \tag{2.6}$$

where $p(x, y)$ represents a pixel in the DSCC, $\bar{x}$ and $\bar{y}$ are the means of x and y coordinates, and u_{mn} is a central moment. For horizontal line detection, we only preserve those DSCCs with either very small sizes ($\max\{a, b\} < T_1$) or large aspect ratios within a specified orientation ($a/b > T_2$ and $\theta \in [-45^\circ, 45^\circ]$). T_1 and T_2 are thresholds determined experimentally. The first condition preserves small DSCCs, which may be parts of a broken line or the touching areas of lines and text; and the second preserves large DSCCs, which are likely to be horizontal line segments. For rule line detection, we need to perform text filtering only in the horizontal direction. Our approach can be extended for vertical line detection, as discussed in the next chapter on known form processing. For vertical line detection, similar filtering conditions exist except for the orientation. Fig. 2.3 shows examples of text filtering. We can see that most text strokes are filtered and the line segments are preserved.

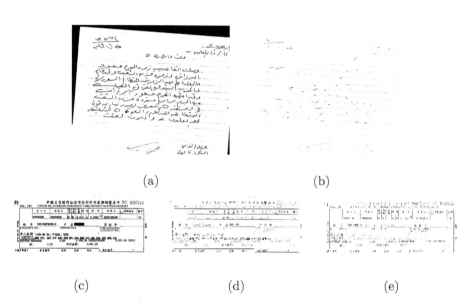

(a)   (b)

(c)   (d)   (e)

Figure 2.3: DSCC-based text filtering. (a) and (b) A document image with rule lines and the corresponding result of text filtering in the horizontal direction. (c) A form document image. (d) and (e) Results of text filtering in the horizontal and vertical directions of (c), respectively.

2.4   HMM-Based Parallel Line Detection

In the following description, we use horizontal line detection as an example to illustrate the proposed method. The extension to vertical line detection is straightforward. After skew correction and text filtering, we perform a horizontal projection and detect lines on the projection profile. A stochastic model, $\mathcal{M}(y_1, y_2, \ldots, y_N)$, is proposed for a group of parallel lines, where N is the number of lines, and $y_i, i = 1, 2, \ldots, N$, is the vertical position of the ith line on the projection profile. The line gap g_i between two neighboring lines i and $i + 1$ is defined as

$$g_i = y_{i+1} - y_i. \tag{2.7}$$

A global image registration method (such as affine transformation or projective transformation) cannot compensate for local distortions introduced in photocopying and scanning. Such local distortions will introduce variations to the vertical line positions y_i's on the horizontal projection profile. Kanungo and Haralick [50] found that the variation of the position of a point is as large as four pixels after removing the global projective deformation. Therefore, the variation of the distance between two points will be within the range [-8, +8] pixels, if the variations of two points are independent. Considering the case that documents may be bent, folded and unfolded, or they may be stored in various environmental conditions (e.g., hot, cold, dry or humid) for years, the local distortions in scanned images may be larger. In our experiments, we found the maximum variation of g_i from its mean value can reach up to 11 pixels. It is hard to model the dependency among the variations of g_i's. As a simplification, in our approach, we do not consider such dependency.

22

Then, it is easy to show that the line positions y_i's form a Markov chain under this simplified assumption. As they are not observable directly, an HMM is more suitable for modeling the projection profile. The line positions can be detected by decoding the HMM.

2.4.1 Hidden Markov Model

The Markov property of a sequence of events is well studied in the literature [51]. Consider a system that stays at one of a set of N distinct states, $S_1, S_2, \ldots, S_N$, at any sampling time t. The system undergoes a change of state according to a set of probabilities associated with the state during the period between two successive sampling times. For a Markov chain (the first order), the probability distribution of q_t only depends on the value of the previous state q_{t-1}

$$P[q_t = S_{k_t} | q_{t-1} = S_{k_{t-1}}, q_{t-2} = S_{k_{t-2}}, \ldots, q_1 = S_{k_1}] = P[q_t = S_{k_t} | q_{t-1} = S_{k_{t-1}}] \quad (2.8)$$

If the state transition probability is independent of time t, then the Markov chain is said to be *homogeneous*

$$P[q_t = S_j | q_{t-1} = S_i] = a_{ij} \quad 1 \leq i, j \leq N \quad (2.9)$$

We can show that line positions $\{Y_i, i = 1, 2, \ldots, N\}$ form a Markov chain, if the variations in line gaps are independent. We use uppercase characters to represent random variables (e.g., Y_i) and lowercase characters to represent the value of the random variables (e.g., y_i).

Theorem 1: Let $Y_i, i = 1, 2, \ldots, N$ be line positions, and $G_i = Y_{i+1} - Y_i, i =$

$1, \ldots, N-1$ be line gaps. If $\{G_i\}$ are independent, then $\{Y_i\}$ form a Markov chain.

$$P(Y_i|Y_1, Y_2, \ldots, Y_{i-1}) = P(Y_i|Y_{i-1}) \qquad (2.10)$$

Proof:

$$
\begin{aligned}
P(Y_i|Y_1, Y_2, \ldots, Y_{i-1}) &= P(G_{i-1} + Y_{i-1}|Y_1, Y_2, \ldots, Y_{i-1}) \\
&= P(G_{i-1}|Y_1, Y_2, \ldots, Y_{i-1}) \\
&= P(G_{i-1}|G_1, G_2, \ldots, G_{i-2}, Y_{i-1}) \qquad (2.11)
\end{aligned}
$$

Since, $\{G_i\}$ are independent, we have

$$
\begin{aligned}
P(Y_i|Y_1, Y_2, \ldots, Y_{i-1}) &= P(G_{i-1}|Y_{i-1}) \\
&= P(G_{i-1} + Y_{i-1}|Y_{i-1}) \\
&= P(Y_i|Y_{i-1}) \qquad (2.12)
\end{aligned}
$$

Therefore, $\{Y_i\}$ form a Markov chain.

In the literature of random process [52], $\{Y_i\}$ is called an independent increment process, which includes several well-known random processes, such as the Brownian motion process and the Poisson process.

In many applications, the actual state sequence is not observable. The resulting model (which is called a hidden Markov model) is a doubly embedded stochastic process with an underlying stochastic process that is not observable, but can be inferred only through another stochastic process that produces the sequence of observations. The elements of a standard discrete HMM are

1) N, the number of the states in the model.

24

2) M, the number of distinct observation symbols per state.

3) $A = \{a_{ij}\}$, the state transition probability matrix.

4) $B = \{b_{ij}\}$, the probability distribution matrix of the observation symbols.

5) π, the initial state distribution.

HMMs can model some 1-D signals well, and have achieved great success in speech [51] and handwriting recognition [6].

In our application, we can observe only the projection profile h_k

$$
P(H_k = h_k | Y_1 = y_1, \ldots, Y_N = y_N) =
\begin{cases}
P(H_k = h_k | \exists i, k = y_i) & \text{A line is on } k \\
P(H_k = h_k | \forall i, k \neq y_i) & \text{No lines are on } k
\end{cases}
\tag{2.13}
$$

Therefore, the projection profile can be modeled with an HMM. A standard HMM is shown in Fig. 2.4a, where S_T and S_B are the states representing top and bottom image borders, $S_{L,i}, i = 1, 2, \ldots, N$, represents lines, and $S_{G,i}, i = 1, 2, \ldots, N-1$, represents the gaps between lines i and $i+1$.

One weakness of conventional HMMs is modeling of the state duration. The inherent duration probability distribution $p_i(d), d = 1, 2, \ldots$, associated with state S_{G_i} is

$$
p_i(d) = (a_{ii})^{d-1}(1 - a_{ii})
\tag{2.14}
$$

where a_{ii} is a self transition probability. The exponential state duration distribution is inappropriate for our applications. Instead we explicitly model the duration distributions. The model with explicit state duration is shown in Fig. 2.4b [1], where

---

[1]It only exactly models lines with one pixel width. To make the model more accurate, alternatively, one can introduce probability of durations to line states. Fortunately, line width does not

25

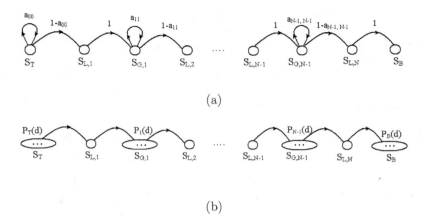

(a)

(b)

Figure 2.4: HMMs for a projection profile. (a) A standard HMM. (b) An HMM with explicit state duration.

the stochastic property of the model is incorporated into the state duration distributions $P_T(d), P_B(d), P_i(d), i = 1, 2, \ldots, N - 1$. For some applications, the quality of the modeling is significantly improved when explicit state duration distributions are used [53].

2.4.2   HMM Parameter Estimation

The major drawback of an explicit duration HMM is that it significantly increases computational costs for model training. With a traditional forward-backward training algorithm (a type of EM algorithm), the re-estimation problem for a variable duration HMM is more difficult than that for a standard HMM [51]. Fortunately, in our case, we can directly derive the HMM parameters from ground-truth since the

vary too much in most applications. Experiments shows that such inaccuracy in modeling does not deteriorate its performance noticeably.

states explicitly correspond to image components. Therefore, the forward-backward training algorithm is not needed. We set duration probabilities of states S_T and S_B to uniform distribution within a range. The duration probabilities of states $S_{G,i}, i = 1, 2, \ldots, N - 1$, is estimated directly from the ground-truth.

The observation comes from the projection profile h_k. The large number of observation symbols would prevent us from estimating the model parameters reliably with limited training samples. There are two methods to reduce the number of parameters of the model. One involves modeling the distribution of the observation as a Gaussian distribution [51], so only the mean and variance of the Gaussian distribution need to be estimated. For known form processing (discussed in the next chapter), we find the projections of a line over multiple form instances can be well modeled as a Gaussian distribution. Another method quantizes the projection profile into several levels. For rule line detection, the image quality varies significantly among different images. The distribution of the observations does not follow a Gauss distribution. Therefore, we quantize h_k into K levels ($K = 5$ in our experiments for rule line detection). The probability of each level is estimated from the ground-truth.

The HMM parameters estimated directly from the ground-truthed data set are not optimal due to the sparseness of the training data. For example, some entries of the line gap distribution do not appear or appear only a few times. Parameter sharing, a technique used in neural networks to train the parameters with limited training samples [54, 55], is used in our approach. For example, we let non-line states $S_T, S_B, S_{G,1}, \ldots, S_{G,N-1}$ share the same observation probability distributions since the observations of these states are the same: the projections of noise and remaining text strokes after filtering. For rule line detection, we further combine all line states into one state, and all non-line states into another state, significantly reducing the parameters of the model. For line gap distribution estimation, we assume the distribution is symmetric around the mean value. Therefore, data smoothing techniques, originally proposed in natural language processing [56], can be used

$$C'(\bar{g}_i + k) = C'(\bar{g}_i - k) = \frac{C(\bar{g}_i + k) + C(\bar{g}_i - k)}{2} \quad k = 1, 2, \ldots \quad (2.15)$$

where $\bar{g}_i$ is the mean value of line gap G_i, $C(k)$ is the number of instances of G_i with value k in the training set, and $C'(k)$ is the smoothed result after imposing symmetric regularization. Finally, we set the empty entries to the minimal value of all non-zero entries. Suppose the maximal variation of the line gap G_i is K. For $k \in [-K, K]$, the final smoothed result is

$$C''(\bar{g}_i + k) = \begin{cases} C'(\bar{g}_i + k) & \text{if } C'(\bar{g}_i + k) \neq 0 \\ \min_{i \in [-K,K], C'(\bar{g}_i+i) \neq 0} C'(\bar{g}_i + i) & \text{if } C'(\bar{g}_i + k) = 0 \end{cases} \quad (2.16)$$

$C'''(k)$ can be converted to probability by normalization.

The ultimate goal of training is to search the optimal HMM parameters to minimize the line detection error. The estimated parameters from the training data can produce reasonable results, but they do not minimize the pre-defined line detection error rate. Generally, the error criterion is a complex function of the model parameters without a closed-form representation. A direct searching algorithm can be used to solve such optimization problems. In our case, the simplex search method proposed by Nelder and Mead [57] is used to minimize the detection error.

28

2.4.3 HMM Decoding

Given the observation sequence $O = h_k, k = 1, 2, \ldots, T$, and the HMM λ, we want to search an optimal state sequence $Q = q_1 q_2 \ldots q_T$ to maximize $P(Q|O, \lambda)$, which is equivalent to maximizing $P(Q, O|\lambda)$. Normally, the Viterbi algorithm, a dynamic programming method, is used to decode HMMs. A matrix v with dimension $T \times (N+1)$ is defined and updated in the Viterbi algorithm, and

$$v(t, n) = \max_{q_1, q_2, \ldots, q_{t-1}} P[q_1, q_2, \ldots, q_t = S_{L,n}, h_1, h_2, \ldots, h_t | \lambda] \qquad (2.17)$$

is the best decoding score at time t, which accounts for the first t observations and ends in state $S_{L,n}$. The sequence $q_1, q_2, \ldots, q_{t-1}$ maximizing the probability in Eq. (2.17) is the best decoding result until time t if we decode state q_t as the nth line.

Suppose the minimal and maximal state durations of states $S_{G,n}$ are δ_{n-} and δ_{n+}, and the durations of S_T and S_B are uniformly distributed in $[0, \delta_T]$ and $[0, \delta_B]$, respectively. The complete procedure of decoding is stated as follows

1. Clear all entries of matrix v.

2. For $1 \leq i \leq \delta_T$, decode the first $i-1$ observations as S_T (the top image border) and observation i as $S_{L,1}$

$$v(i, 1) = \frac{1}{\delta_T + 1} P(h_i | q_i = S_{L,1}) \prod_{j=1}^{i-1} P(h_j | q_j = S_T), \qquad (2.18)$$

where $P(h_i | q_i = S_{L,1})$ is the probability of observing h_i if the system enters state $S_{L,1}$ at time i; $\prod_{j=1}^{i-1} P(h_j | q_j = S_T)$ is the probability of observing the first $i-1$ observations if the system stays at state S_T during the time period from 1 to $i-1$; and $\frac{1}{\delta_T+1}$ is the probability of the model staying at S_T for $i-1$ consecutive periods.

3. Set $t=1$.

4. For $n=1$ to N

   For $j=\delta_{n-}$ to δ_{n+}

$$v'(t+j,n) = v(t,n)P_n(j)P(h_{t+j}|q_{t+j}=S_{L,n+1}) \prod_{k=t+1}^{t+j-1} P(h_k|q_k=S_{G,n}) \quad (2.19)$$

$$v(t+j,n) = \max\{v(t+j,n),\ v'(t+j,n)\} \quad (2.20)$$

   End loop of j

   End loop of n

   Here $P_n(j)$ is the probability of staying at state $S_{G,n}$ with j consecutive times; $P(h_{t+j}|q_{t+j}=S_{L,n+1})$ is the probability of observing observation h_{t+j} if the system enters $S_{L,n+1}$ at time $t+j$, which corresponds to a new line; and $\prod_{k=t+1}^{t+j-1} P(h_k|q_k=S_{G,n})$ is the probability of observing sequence h_{t+1} to h_{t+j-1} if the system stays at state $S_{G,n}$ during this time period, which corresponds to a line gap. Eq. (2.20) updates the optimal partial detection result.

5. If $t>T-\delta_B$, decode the following sequence as the bottom image border.

$$v'(T,N+1) = v(t,N)\frac{1}{\delta_B+1}\prod_{k=t+1}^{T} P(h_k|q_k=S_B) \quad (2.21)$$

$$v(T,N+1) = \max\{v(T,N+1),\ v'(T,N+1)\} \quad (2.22)$$

6. If $t<T$, then $t=t+1$, and go to step 4.

30

For each t, the algorithm remembers the best decoding path until time t. After decoding,

$$v(T, N + 1) = \max_{Q} P(Q, O|\lambda) \qquad (2.23)$$

is the probability of detecting lines given the model, which can be regarded as detection confidence. The sequence $q_1, q_2, \ldots, q_T$ which achieves $v(T, N + 1)$ is the optimal decoding result.

2.4.4 Polyline Representation

After identifying the vertical position of a line, we then need to detect the left and right end points by grouping the broken line segments together. For each detected line, those DSCCs within 10 pixels distance to the detected line are merged [1].

An ideal straight line can be represented with two parameters a and b as $y = a \times x + b$. Practically, a real line is represented with points $(x_i, y_i), i = 1, 2, \ldots, n$. The parameters a and b can be estimated based on the minimum mean squared error criterion (MMSE)

$$\bar{x} = \sum_{i=1}^{n} x_i/n,$$

$$\bar{y} = \sum_{i=1}^{n} y_i/n,$$

$$a = \frac{\sum_{i=1}^{n}(x - \bar{x})(y - \bar{y})}{\sum_{i=1}^{n}(x - \bar{x})^2},$$

$$b = \bar{y} - a \times \bar{x}. \qquad (2.24)$$

For most straight lines, this approximation is good enough. However, due to the distortions introduced by photocopying and scanning, some lines are cursive and cannot be well represented by two end points. In this case, a polyline representation is used as follows:

1. Calculate the average approximation error of a line

$$\delta y_i \;=\; |y_i - a \times x_i - b|, \qquad (2.25)$$

$$e \;=\; \sum_{i=1}^{n} \delta y_i / n. \qquad (2.26)$$

2. If e is smaller than the average line width (often two to four pixels), keep it with two end points representation and exit.

3. Otherwise, split the whole line into two segments from the middle and estimate the line parameters a and b for each segment respectively, as described in Eq. (2.24).

4. For each segment, go to step 1 and repeat.

A polyline is described as a sequence of vertices $(P_1, P_2, \ldots, P_m)$. Two or three segments are sufficient to represent most lines in our following experiments.

2.5 Application to Rule Line Detection

In this section we use the proposed method to detect severely broken rule lines. In this application, the number of lines is unknown, and the vertical line gaps may vary in different images due to the different styles used by rule-lined paper or different scanning resolutions. However, the length of lines and the vertical line gaps are roughly consistent in the same document image.

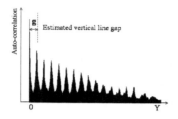

Figure 2.5: Vertical line gap estimation for rule line detection, based on the auto-correlation of the projection profile.

2.5.1 Vertical Line Gap Estimation

We need to estimate the average vertical line gap from the input image. Since the line gaps between neighboring lines are roughly the same, the horizontal projection of rule lines is a periodic signal (the period is the average vertical line gap $\bar{g}$). We use an auto-correlation-based approach to estimating the period of the projection. The auto-correlation of a signal x, with n samples $x(1), x(2), \cdots, x(n)$, is defined as

$$R(l) = \sum_{i=1}^{n-l} x(i)x(i+l) \quad l = 0, 1, \ldots, n-1 \tag{2.27}$$

The distance between the first two peaks of the auto-correlation is taken as the vertical line gap, as shown in Fig. 2.5.

2.5.2 A Simplified Model

In order to reduce the complexity of the model (the number of states and parameters), we further simplify it by considering the special properties of rule lines. Since the vertical line gaps and the lengths of rule lines are roughly consistent in the same document image, we can merge states $S_{G,i}, i = 1, 2, \ldots, N-1$, into one state S_G, and $S_{L,i}, i = 1, 2, \ldots, N$, into another state S_L. Fig. 2.6 shows the simplified model.

33

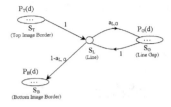

Figure 2.6: A simplified HMM for rule line detection.

State merging reduces the number of parameters significantly. Another advantage of such simplification is that we do not need to know explicitly the number of lines on a document.

2.5.3 Parameter Estimation

In our data set, the quality of different images varies significantly as does the quality of rule lines on the same image. Therefore, we cannot use the Gaussian distribution to model the projections of rule lines (the Gaussian mixture distributions may be a good approximation). Instead, we quantize the observation into several levels and estimate the probability of each quantized level directly from the ground-truthed data set. Peaks on the projection profile have particular significance for line detection. Therefore, we first set all non-peaks on the profile to zero, then quantize the peaks on the projection profile into four levels using the following quantization thresholds: $w/16, w/8$, and $w/4$, where w is the image width. The observation probability distribution matrix B, estimated from the training set containing 100 documents, is listed in Table 2.1. We let states S_T, S_B, and S_G, whose observations are the projection of text or noise, share the same observation distribution. We ob-

Table 2.1: Observation probability distribution matrix B estimated from the training set containing 100 documents

	0 Non-peak	1 $(0, \frac{w}{16}]$	2 $(\frac{w}{16}, \frac{w}{8}]$	3 $(\frac{w}{8}, \frac{w}{4}]$	4 $(\frac{w}{4}, w]$
S_L	106 (4.7%)	246 (10.8%)	378 (16.6%)	1,051 (46.2%)	493 (21.7%)
S_T, S_B, S_G	191,086 (98.8%)	2,052 (1.1%)	170 (0.1%)	58 (0.03%)	15 (0.008%)

served that (1) due to the severe brokenness, the horizontal projections of about 80% of rule lines are less than 1/4 of the image width; (2) 4.7% of rule lines do not form peaks; and (3) the peaks with small heights are more likely formed by text strokes or noise (2,052 instances) rather than by rule lines (246 instances). Therefore, we need to use high level contextual information to achieve reasonable detection results for these severely broken lines.

We set duration probability of states S_T and S_B to the uniform distribution on $[0, \bar{g}-1]$. The duration probability of state S_G is estimated directly from the ground-truth with the approach described in Section 2.4.2. With all these settings, the rule line detection accuracy on the training set is about 95.6%. For comparison, the accuracy is only 91.7% if we use the Gaussian distribution for approximation. Since the parameters estimated from the training data are not optimal for the ultimate detection error criterion, the simplex method proposed by Nelder and Mead [57] is used to search the optimal parameter set which minimizes the detection error. Among the parameters of our model, we optimize only the observation probability matrix B. Experiments show the detection accuracy increases to 97.3% on the training set after optimization.

2.5.4 Examples

HMM decoding may detect extra lines on the top or bottom image borders. To reduce the false alarm rate, we remove lines with less than 50 black pixels. Fig. 2.1c shows the model-based line detection result for a rule-lined document. Compared with Fig. 2.1b, we can see that, with contextual information, the result is significantly improved. Our model-based method is robust even when the input images do not follow the model exactly. Fig. 2.7a shows an example: two pages are overlapped during scanning. Our algorithm still detects all rule lines correctly. In Fig. 2.7c, we remove 35 rows of the image (about half of the average vertical line gap of this document). The variation of the line gap is out of the range allowed by the model. The corresponding detection result is shown in Fig. 2.7d, with only one line missed due to the anomalous vertical line gap.

2.6 Experiments

In this section, we present our evaluation metrics, quantitatively evaluate the robustness of our line detection algorithm, and compare it with several non-model-based algorithms.

2.6.1 Line Detection Evaluation Protocol

Line detection accuracy can be evaluated at the pixel and line levels [58]. At the pixel level we compare the difference of the pixels between ground-truth and detected lines. It is straightforward and objective, but ground-truthing at the pixel level is

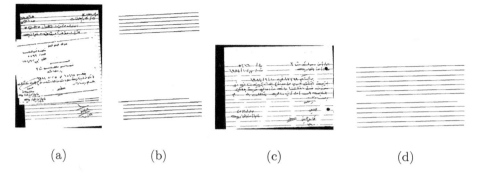

| (a) | (b) | (c) | (d) |

Figure 2.7: Robust test for rule line detection. The image in (a) has enormous observations on the projection profile (several lines missed because two pages are overlapped during scanning). (c) shows a document with enormous line gaps (35 image rows removed manually in the middle). (b) and (d) show the corresponding line detection results.

extremely expensive when lines are broken, distorted, and overlapped with text. Therefore, we evaluate the algorithm at the line level. Our evaluation metric is based on the Hausdorff distance. The Hausdorff distance between two point sets is

$$H(A, B) = \max\{h(A, B), h(B, A)\} \tag{2.28}$$

where

$$h(A, B) = \max_{a \in A} \min_{b \in B} ||a - b|| \tag{2.29}$$

and $||.||$ is an underlying norm (e.g., the L_2 or Euclidean distance). The function $h(A, B)$ is called the *directed* Hausdorff distance from A to B. It identifies the point $a \in A$ that is the farthest from any point of B and measures the distance from a to its nearest neighbor in B [59]. The direct computation method for the Hausdorff distance is time consuming, but, for polyline representation, the Hausdorff distance

37

can be easily calculated. Suppose polylines A and B are represented as a sequence of vertices $(A_1, A_2, \ldots, A_m)$ and $(B_1, B_2, \ldots, B_n)$ respectively, then the Hausdorff distance is simplified as

$$H(A, B) = \max\{H'(A, B), e(A, B)\} \tag{2.30}$$

where

$$H'(A, B) = \max\{D_{A1}, D_{A2}, \ldots, D_{Am}, D_{B1}, D_{B2}, \ldots, D_{Bn}\} \tag{2.31}$$

$$e(A, B) = \max\{||A_1 - B_1||, ||A_m - B_n||\} \tag{2.32}$$

D_{Ai} is the perpendicular distance from A_i to polyline B, and D_{Bi} is the perpendicular distance from B_i to polyline A, as shown in Fig. 2.8. $H'(A, B)$ in Eq. (2.31) is the *perpendicular distance* between two polylines A and B, which evaluates the accuracy in determining the vertical location of a horizontal line and the horizontal location of a vertical line. $||A_i - B_j||$ is the Euclidean distance between points A_i and B_j. Suppose the vertices of a polyline are sorted from left to right for a horizontal line, and top to bottom for a vertical line. Then $||A_1 - B_1||$ and $||A_m - B_n||$ are the *end point determination errors*. Hausdorff distance $H(A, B)$ in Eq. (2.30) combines the perpendicular distance and end point determination errors into one metric.

For severely broken lines, however, it is hard to define the end points exactly. Therefore, we prefer to use two separate metrics: the perpendicular distance and end point determination error for evaluation, instead of a combined Hausdorff distance.

The end point determination error is an absolute value. As a supplemental

Figure 2.8: Hausdorff distance between two polylines.

metric, the overlap rate of polylines A and B

$$o(A, B) = \frac{\min\{A_m, B_n\} - \max\{A_1, B_1\}}{\max\{A_m, B_n\} - \min\{A_1, B_1\}} \tag{2.33}$$

is defined to evaluate the relative end point determination error.

As suggested in [60], if a detected line is within no more than five pixels to a ground-truthed line in the perpendicular direction, it is said to be correctly detected. If the perpendicular distance is larger than five pixels and no more than ten pixels, it is said to be partially correct. Splitting and merging errors are all assigned as partially correct too.

2.6.2   Quantitative Evaluation for Rule Line Detection

We obtained 168 Arabic document images with a total of 3,870 ground-truthed rule lines, most of which are severely broken. We use 100 images to train the HMM, and the remaining 68 images as the test set. The detection results on the test set are shown in the last row of Table 2.2. On the test set, 96.8% of lines are detected correctly, with only two lines missed. The false alarm rate is 2.3%. Most of the false alarms are caused by the inconsistency between the detector and the subjective judgment of the ground-truther when lines are severely broken. For correctly detected lines, we evaluate the end point determination accuracy using

39

the end point determination error and overlap rate defined in Eq. (2.32) and (2.33) respectively. The average end point determination error is six pixels and the overlap rate is 99.1%.

We compared our model-based line detection algorithm with other non-model-based line detection algorithms: the Hough transform method [9], the projection method [10], and the DSCC method [1]. Table 2.2 shows the line detection results on the test set with different algorithms. The results of the Hough transform and projection methods listed in the table are tested on the images after text filtering. The projections of lines often fail to form peaks on the projection profile, if lines overlap with text or they are severely broken. Text filtering helps lines to form peaks on the projection profile, therefore increases the detection rate. On this data set, under roughly the same false alarm rate, the detection accuracy increases from 73% on raw images to 82% on text-filtered images. For either projection or Hough transform methods, only those peaks with values larger than a threshold are picked as line positions. With a small threshold, we can detect more lines, but the false alarm rate is high. Increasing the threshold will reduce the false alarm rate, but increase the mis-detection rate. We selected the threshold to make the false alarm rate roughly equal the mis-detection rate. To reduce the false alarm rate of the Hough transform method further, we restrict the search range of θ to $[-1^\circ, 1^\circ]$ after skew correction. For the DSCC method, we restrict the merging direction to the horizontal direction. As expected, our model-based method achieved much better results in both accuracy and false alarm rate, due to the use of high level constraints between neighboring lines.

Table 2.2: Comparison of our model-based method with other methods on the test set for rule line detection (there are a total of 1,596 ground-truthed lines).

	Detected	Correct	Partial Correct	Missed	False Alarm
Hough Transform	1,588	1,299 (81.4%)	60 (3.8%)	237 (14.9%)	229 (14.4%)
Projection Method	1,577	1,310 (82.1%)	112 (7.0%)	174 (10.9%)	155 (9.7%)
DSCC	2,162	1,398 (87.6%)	118 (7.4%)	80 (5.0%)	646 (40.5%)
Our Model-Based Method	1,631	**1,545 (96.8%)**	**49 (3.0%)**	**2 (0.1%)**	**37 (2.3%)**

2.7 Summary and Future Work

We present a novel approach to detect severely broken rule lines in documents. Our method is based on a stochastic model to incorporate high level constraints into a general line detection algorithm. Instead of detecting lines individually, we use the Viterbi algorithm to detect all parallel lines simultaneously. Our method can detect 96.8% of the severely broken rule lines in the Arabic database we collected. Some challenging examples demonstrated the robustness of our approach.

After detection, rule lines must be removed before further document processing. Fig. 2.1d shows the document image after rule line removal. The result is reasonable; when text strokes overlap with lines, some parts of text strokes may be removed erroneously. A more robust method should be developed to improve the line removal results.

Chapter 3

Known Form Processing

3.1 Introduction

Millions of form documents, such as tax return forms, health insurance forms, airline vouchers, checks, and bank slips, are processed everyday [15, 16, 61, 62, 63, 19]. Processing of such documents can be categorized as *unknown* and *known* form processing [16]. Unknown form processing assumes no a priori knowledge from the input forms, and extracts all information based on low level image analysis. Errors are expected and user assistance is required. Known form processing, on the other hand, is designed to process a pre-defined set of forms, where a priori information can be stored as templates in the database to guide later processing. It is widely used in banks, post offices, and tax offices where the types of forms are most often pre-defined. For an input form, the system first selects the template that it matches best (*form identification*), then extracts anchors (such as specific marks and form frame lines) for registration to compensate for variations produced by scanning (e.g., rotation, translation, scaling, and local nonlinear distortions) [1]. Finally, the identified template is used to guide the system to recognize fields of interest on the form

---

[1]If preprinted content (fixed part) dominates user-filled information (variant part), general image registration methods (e.g., correlation-based methods) can be used for form registration without detecting any lines or landmarks.

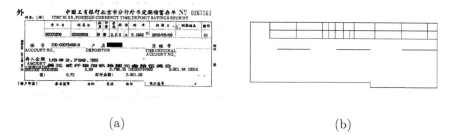

(a)   (b)

Figure 3.1: (a) An example deposit form of the Industrial and Commercial Bank of China. There are two groups for parallel lines on this form, one horizontal and the other vertical. (b) Line detection result using our model-based approach.

(different OCR engines may be used for different fields), and output the recognition results to a database. Although special anchors may be available to facilitate form identification and registration for specially designed forms, more general approaches use features related explicitly or implicitly to frame lines, such as the frame lines themselves [15, 16, 61, 62], form cells [63], and intersections of frame lines [19]. Robust detection of frame lines is crucial to these approaches. In the previous chapter, we proposed a model-based parallel line detection algorithm using hidden Markov models (HMM). In this chapter, we apply it for known form processing. As shown in Fig. 3.1a, generally there are two groups of parallel lines (one horizontal and the other vertical) on a form, so we use two HMMs to detect the horizontal and vertical lines separately. The detected lines can be used for registration. Our algorithm can be extended for form identification too, so in our approach, a unified framework solves both important tasks in known form processing.

For known form processing, we need to not only detect lines reliably, but also find the correspondence between the detected lines and those stored in the form template [15, 16]. The method proposed by Tang et al. [15] assumes there is only one anchor line in a pre-defined region, which can be distinguished easily from other lines. The application of this method is restricted. Considering false alarms and mis-detections, the correspondence problem is not trivial. Cesarini et al. [16] proposed a hypothesis and verification paradigm as a solution. For a detected line in a pre-define region, several hypotheses are generated about correspondence between the line and those in the template. Under each hypothesis, the rough positions of other lines can be determined, then the system searches the expected lines to verify the hypothesis. The output of the verification module is binary: success or failure. All lines used for registration should be detected to achieve a consistent solution, so it is not robust to line degradations. Both methods need an initial region to detect the first anchor line, and only a subset of lines are used for registration. In our approach, we use all lines for registration, but we do not perform binary assertion during HMM decoding. Instead, we measure the probability of a projection to be generated by a line. The optimal detection results are achieved by the Viterbi algorithm. The degradation of a few lines may not deteriorate the performance. Another advantage of our approach is that the detection and correspondence problems are solved simultaneously. After HMM decoding, the correspondence between the detected lines and those in the form template (or the model) is achieved automatically.

The remainder of this chapter is organized as following. In Section 3.2, we apply our general model-based line detection algorithm for form frame line detection. The quantitative evaluation of the robustness of our approach is presented in Section 3.3. Our approach can be extended for form identification, which is described

44

in Section 3.4. This chapter ends with a brief summary in Section 3.5.

3.2   Form Frame Line Detection

The application of the algorithm to known form processing is straightforward. Generally, a collection of horizontal and vertical parallel lines exists on a form, so we use two HMMs to detect the horizontal and vertical lines separately. To apply the algorithm, we need to estimate two sets of parameters: (1) The distribution of observation symbols of each state; and (2) The state duration probabilities of each gap state.

3.2.1   Estimation of the Distributions of Observation Symbols

In our case, the observation symbols are the projection profile, which has the range of $[0, w]$ for a horizontal projection (where w is the width of the image). As we stated previously, a large number of observation symbols would cause difficulties in reliably estimating the distributions with limited training samples. With some assumptions, we can show that using a Gaussian distribution to model projections of a line over multiple form instances is appropriate. In a widely used stochastic document image degradation model [64], a white (black) pixel is randomly selected and flipped to black (white). The projection is the summation of all black pixels on the line

$$h = \sum_{i=1}^{M} a_i \qquad (3.1)$$

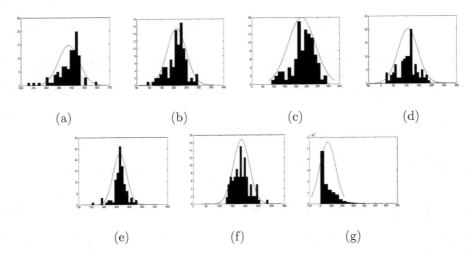

(a) (b) (c) (d)

(e) (f) (g)

Figure 3.2: The distributions of the observation symbols (horizontal projections) for 100 scanned instances of a bank deposit form. One example of the form is shown in Fig. 3.1a. There are six horizontal lines on the form. (a) to (f) The histograms of the projections of six horizontal lines respectively. (g) The histogram of the projections of the non-line states.

where

$$
a_i = \begin{cases} 1 & \text{if black pixel } i \text{ is preserved} \\ 0 & \text{if black pixel } i \text{ is flipped to white during degradation} \end{cases} \tag{3.2}
$$

Under white Gaussian noise (a widely used model for degradation), a_i follows a Bernoulli distribution: $a_i \sim Bernoulli(\rho)$, where ρ is the probability for a black pixel to be lost. Consequently, h follows a binomial distribution $Bin(\rho, M)$

$$
P(h) = \binom{M}{h} \rho^{M-h}(1-\rho)^h. \tag{3.3}
$$

According to the central limit law, if M is large enough (or if the line is long enough), then the distribution of random variable h converges to a Gaussian distribution [52]

$$
\lim_{M \to \infty} \frac{h - E[h]}{\sqrt{M\rho(1-\rho)}} \longrightarrow \mathcal{N}(0, 1) \quad \text{in distribution.} \tag{3.4}
$$

In known form processing, a set of forms are captured with similar imaging conditions. Therefore, ρ is roughly constant for each form in the set. A Gaussian distribution is a good approximation for the projections of a line over multiple form instances. The mean and variance of the Gaussian distribution can be estimated from the ground-truth. Figs. 3.2a to 3.2f show the distributions of the projections of all six horizontal lines on a set of bank deposit forms with one instance shown in Fig. 3.1a. The histogram is generated over 100 form samples. We can see that the Gaussian distribution is a good approximation. For non-line states, the approximation is not good enough, since a projection is always no less than zero (an exponential distribution may be more suitable), as shown in Fig. 3.2g. We found in the experiments that the effect of this approximation error is negligible for the final line detection result.

Table 3.1: The distribution of the line gap between the first and second horizontal lines on a bank deposit form. The average is 94 pixels. The row of *distance* lists the difference to the average value.

Distance	-9	-8	-7	-6	-5	-4	-3	-2	-1	0	1	2	3	4	5	6	7	8	9
Raw Occurrence	1	0	2	0	0	3	5	12	18	24	16	6	8	4	0	1	0	0	0
Symmetric Regularization	.5	0	1	.5	0	3.5	6.5	9	17	24	17	9	6.5	3.5	0	.5	1	0	.5
Zero-Occurrence Smoothing	.5	.5	1	.5	.5	3.5	6.5	9	17	24	17	9	6.5	3.5	.5	.5	1	.5	.5

3.2.2 Estimation of the State Durations

The state duration of $S_{G,i}, i = 1, 2, \ldots, N-1$, represents the line gap between lines i and $i+1$, which can be estimated from the ground-truth. Table 3.1 shows the distribution of the gap between the first and second horizontal lines on the bank deposit form in our database of 100 samples. The average value of the gap is 94 pixels. The row of *distance* lists the difference to the average value. The row of *raw occurrence* shows the number of occurrences in which the gap takes a specific value. We can see that the variation is from -9 pixels to 6 pixels, and the distribution is roughly symmetric around the average value. Due to the sparse-data problem, some entries within the range of [-9, 6] are not observed in the training set, which will deteriorate the performance. Therefore, data smoothing is used. The row of *symmetric regularization* is the result after we impose the symmetry. Lastly, we set the zero entries to the minimal value of all non-zero entries, as shown in the row of *zero-occurrence smoothing*. After data smoothing, the distribution $P_1(d)$ can be estimated by normalization. Similarly, we can get the distributions of other line gaps.

3.2.3 Decoding

After estimating parameters, we use the Viterbi algorithm to decode the observation. Fig. 3.3 shows the decoding results of the Viterbi algorithm on the horizontal and vertical projection profiles of the bank deposit form (Fig. 3.1a). The locations found by the Viterbi algorithm are labeled with squares. We can see that instead of picking the highest peaks as detected lines in the projection methods [10, 13], our approach outputs the line positions most compatible with the model.

After detecting the horizontal and vertical lines, the method described in the previous chapter can be used to determine the end points of the lines. However, if a line is severely degraded, the end points cannot be determined accurately. For many forms, the intersections of horizontal and vertical lines can be used to determine the end points. Sometimes, several lines may lie on the same line, for example, three dashed lines in the middle of the form as shown in Fig. 3.1a. Our HMM-based method can handle this special case without difficulty. In this example, the vertical line gaps between dashed lines are zero. They share the same horizontal projection. The Viterbi algorithm gives the vertical position of these lines. The left and right end points are determined using their position relative to the intersections of horizontal and vertical lines. In this case, horizontal and vertical lines should be extended to get the intersection points. Fig. 3.1b shows the model-based line detection result. We can see our method can detect the short lines which may not form peaks on the projection profile (especially for the two shortest vertical lines), which are most likely missed by other methods such as the Hough transform or projection methods.

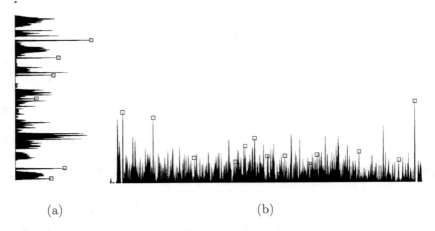

(a)   (b)

Figure 3.3: The lines detected after decoding the HMMs using the Viterbi algorithm on the horizontal (a) and vertical (b) projection profiles of the bank deposit form. The original form is shown in Fig. 3.1a. The locations picked up by the Viterbi algorithm are labeled with squares.

Our method outputs the exact number of lines indicated by the model without false alarms. Fig. 3.4 shows two more examples of an export registration form used by the Customs Bureau of China and a portion of a US income tax form.

3.3   Experiments for Form Frame Line Detection

To evaluate the algorithm for known form processing, we collected 100 bank deposit forms. In this experiment, we did not evaluate the accuracy of form registration directly. The accuracy of form registration depends on which deformation model (global affine transformation or more flexible local deformation) is used to transform the input form to the prototype form. Since the detected lines are used for both form identification (discussed in the next section) and registration, we evaluate the

50

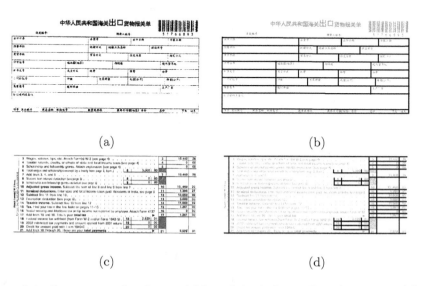

(a)   (b)

(c)   (d)

Figure 3.4: Some examples for model-based form frame line detection. (a) and
(b) An export registration form used by the Customs Bureau of China and the
corresponding line detection result. (c) and (d) A portion of a US income tax form
and the corresponding line detection result. The detected lines are shown in black
and overlay with the original documents.

line detection accuracy.

The experiment demonstrated that one training sample can achieve reasonable results if image quality is good. We selected the first image for training. The real value of the projection of a line in this training sample is taken as the mean of the observation random variable of the corresponding line state. The variance of the observation random variable of a line state is set as 20% of its mean. The distribution of line gaps is set within the range of [-10, 10] pixels around its real value in this sample. We tested it on the remaining 99 form images. The last row in Table 3.2 shows the result, using the evaluation metrics defined in the previous chapter. All lines are detected without any false alarms. Only four lines are detected with large location errors. For comparison, Table 3.2 shows the detection results of other algorithms. Both Hough transform and projection methods need a threshold, the minimum pixels on a line, to reduce the false alarm rate. To avoid using an arbitrarily threshold, we selected the first six longest horizontal lines and 14 longest vertical lines as the detection results for both the Hough transform and projection methods. Our algorithm clearly outperforms all three general line detection methods in both mis-detection and false alarm rates.

In the following experiments, we tested the robustness of our method under different scanning resolutions, scanning binarization thresholds, and synthesized image degradations. Generally, the more severe the degradation, the more accurate the model should be in order to detect lines correctly. Therefore, in the following experiments, we increased the number of training samples. We randomly selected 50 forms for training, and used the remaining 50 forms for testing. Fig. 3.5a shows

Table 3.2: Comparison of our model-based method with other methods for known form processing (there are a total of 1,980 ground-truthed lines).

	Detected	Correct	Partial Correct	Missed	False Alarm
Hough Transform	1,980	1,675 (84.6%)	8 (0.4%)	297 (15.0%)	297 (15.0%)
Projection Method	1,980	1,745 (88.1%)	15 (0.8%)	223 (11.3%)	220 (11.1%)
DSCC	2,032	1,803 (91.1%)	125 (6.3%)	175 (8.8%)	104 (5.3%)
Our Model-Based Method	1,980	**1,976 (99.8%)**	**4 (0.2%)**	**0 (0.0%)**	**0 (0.0%)**

the line detection accuracy under different scanning resolutions. As we can see, the performance of the algorithm stays consistently high under a wide range of scanning resolutions from 75 dpi to 600 dpi. The line width varies from about one pixel under 75 dpi resolution to 10 pixels under 600 dpi. Though our model does not include the duration of line states, this inaccuracy in modeling has a negligible effect on its performance.

In the next experiment, we fixed the scanning resolution to 300 dpi and used different binarization thresholds during scanning. If the threshold is too small, the lines are severely broken as shown in Fig. 3.6a (with the threshold of 40). If the threshold is too large, text and lines are smeared together, as shown in Fig. 3.6c (with the threshold of 240). As shown in Fig. 3.6b and d, our algorithm can still detect lines correctly under such extreme conditions. The quantitative evaluation result is shown in Fig. 3.5b. The curve labeled with ∘ in the figure shows the detection accuracy when the training set and test set are scanned with the same binarization threshold. In most applications, the test set may have different characteristics with the training set. The curve labeled with + in Fig. 3.5b shows the detection accuracy

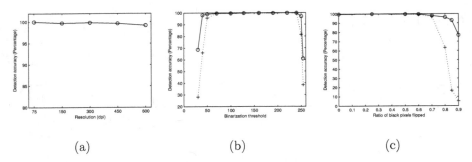

(a)   (b)   (c)

Figure 3.5: Robustness testing. The curves labeled with ○ shows the detection accuracy under the condition where the test set has the same degradation level with the training set. Curves labeled with + are the results when the test set and the training set have different degradation levels. (a) Scanning resolution. (b) Binarization threshold. (c) Synthesized degradation.

using the HMM trained on the training set scanned with a binarization threshold of 128. As we can see, good results are achieved in a wide range even though the test set has different characteristics with the training set.

Synthesized data are often used to test an algorithm because it can directly control the image quality of the test samples. In the following experiments, we selected the data set with good image quality (scanned with 300 dpi and the binarization threshold of 128), and randomly flipped a certain ratio of black pixels on lines to white, keeping all pixels on text unchanged. Figs. 3.7a and 3.7c show the degraded images with 50% and 90% black pixels on lines flipped to white, respectively. As shown in Fig. 3.7b, the line detection result is perfect even if half the black pixels are flipped. Fig. 3.7d shows that the horizontal lines are still detected correctly even when 90% black pixels are flipped, but the vertical lines are misdetected. Fig. 3.5c

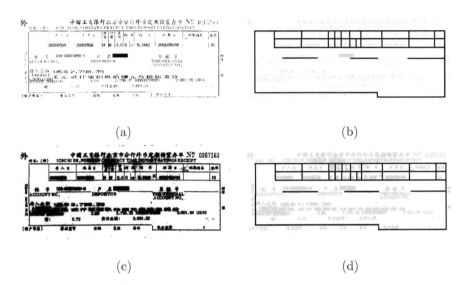

(a)   (b)

(c)   (d)

Figure 3.6: Scanned form documents with different binarization thresholds and the corresponding line detection results. (a) and (c) Scanned images under thresholds of 40 and 240 respectively. (b) and (d) are corresponding line detection results of (a) and (c). The detected lines are shown in black and overlay with the original documents.

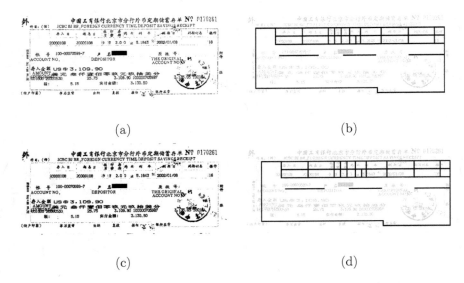

(a)

(b)

(c)

(d)

Figure 3.7: Degraded form documents and the corresponding line detection results. Red lines drawn on original images indicate the detected lines. (a) About 50% of the pixels of the lines are flipped. (c) about 90% of pixels of the lines are flipped. (b) and (d) are corresponding line detection results to (a) and (c).

shows the detection accuracy versus degradation level on the test set. The curve labeled with ○ shows the results when the test sets have the same degradation level with the training sets. We can see that our method is very robust. It maintains good results with accuracy of 96.2% even when 80% black pixels of lines are flipped. The curve labeled with + shows the accuracy on the test sets using the HMM model trained on samples with the degradation level of 50%. Almost the same accuracy is achieved until 70% black pixels on lines are flipped. After that, it breaks down faster.

3.4 Form Identification

Our line detection algorithm can be extended to form identification. Suppose there are n form templates $\lambda_1, \lambda_2, \ldots, \lambda_n$. According to the Bayesian rule, $\hat{\lambda}$ that maximizes the posteriori probability is selected as the template for the input form

$$\hat{\lambda} = arg \max_{\lambda_i} P(\lambda_i|O) = arg \max_{\lambda_i} P(O|\lambda_i)P(\lambda_i). \tag{3.5}$$

Here, O is the observation (the projection profile in our method). $P(\lambda_i)$ is the priori probability of form template λ_i. $P(O|\lambda_i)$ is the probability of observing the sequence of observations given the model λ_i, which can be calculated efficiently with the forward algorithm [51].

Although a Gaussian distribution is a good approximation for observations of a line state, it is not good enough to approximate the observations of a non-line state, as shown in Fig. 3.2g. As demonstrated experimentally, such approximation error does not affect the line detection results noticeably. However, it will make calculating the probability $P(O|\lambda)$ in Eq. (3.5) un-reliable, since the observations are dominated by non-line states. Alternatively, given a form model λ, we first detect lines under the model. Suppose, $h_{L1}, h_{L2}, \ldots, h_{LN}$ are decoded as observations of line states, and $g_1, g_2, \ldots, g_{N-1}$ are line gaps, the probability of the input form sample belonging to the model λ is approximated as

$$Q(O|\lambda) = \sqrt[N]{\Pi_{i=1}^{N} P(h_{Li}|\lambda) \Pi_{i=1}^{N-1} P(g_i|\lambda)} \tag{3.6}$$

In the above equation, we omit the observations of non-line states. The model with the highest probability is selected as the final form identification result.

We test the proposed method on the NIST Structured Forms Reference Set, *NIST Special Database* 2 [65]. The data set consists of 5,590 pages of binary, black-and-white images of synthesized documents. The documents in this database are 12 different tax forms from the IRS 1040 Package X for the year 1988. These include Forms 1040, 2106, 2441, 4562, and 6251 together with Schedules A, B, C, D, E, F, and SE. Eight of these forms contain two pages or form faces for a total of 20 different form faces represented in the database. The number of samples of each form face varies from 59 to 900. The first 20 samples of each form face are used for training and the rest for testing. The form identification results are perfect with an accuracy of 100%.

3.5 Summary and Future Work

In this chapter, we applied our general model-based line detection algorithm to known form processing. There are two tasks in known form processing: form identification and form registration. These two tasks can be solved in one unified framework by extending our model-based line detection algorithm. Our approach is robust under a wide range of scanning resolutions, binarization thresholds, and synthesized degradation levels, as demonstrated experimentally. A further improvement of the proposed work may use the Gaussian mixture distributions or the exponential distribution to replace the simple Gaussian distribution to model the observations of non-line states.

Chapter 4

Handwriting Identification in Noisy Document Images

4.1 Introduction

Handwriting often combines with machine printed text. Handwriting in a machine printed document often indicates corrections, additions, or other supplemental information that should be treated differently from the main content. The segmentation and recognition techniques requested for machine printed and handwritten text differ significantly. Therefore, identification of handwriting from machine printed text is important for the following document image analysis.

Handwriting/machine printed text discrimination can be performed at different levels, such as the text line [19, 20, 21, 22], word [23], or character level [24, 25]. Special consideration must be given to the size of the region being segmented before performing any classification. We call the smallest unit for classification a *pattern unit*. If the unit is too small, the information contained in it may not be sufficient for classification. If it is too large, however, different types of components may be mixed in the same region. In previous work [25] we conducted a performance evaluation for the classification accuracy of machine printed text and handwriting at the character, word, and zone levels. Experiments show that a reliable classification can be achieved at the word level. We therefore segment images at the the word level, then perform classification.

The data set we are processing is noisy, which makes for a more challenging problem. Most document enhancement algorithms can remove large noise components (e.g., marginal black strips) with some simple rules [17, 18], and small noise components (e.g., pepper-and-salt noise) with morphological operations. However, noise components with a compatible size to printed words cannot be easily removed. In our approach, we treat noise as a distinguished class and model it based on selected features. We treat the problem as a three-class (machine printed text, handwriting, and noise) identification problem.

In practice mis-classification happens in an overlapping feature space. This holds especially true for handwriting and noise. To deal with this problem, we exploit contextual information in post-processing and refine the classification. Contextual information helps improve classification accuracy. Many OCR systems use it, and its effectiveness has been demonstrated in previous work [66, 67]. The key is to model the statistical dependency among neighboring components. An OCR system outputs a text stream that is one-dimensional. Therefore, an N-gram language model, based on an Nth order 1-D Markov chain, effectively models the context. With assistance from a dictionary, the N-gram approach can correct most recognition errors. Images, however, are two-dimensional. Generally, 2-D signals are not causal, and it is much harder to model the dependency among neighboring components in an image. Among the image models studied so far, Markov random fields (MRF) have been widely studied and successfully used in many applications [68]. MRFs are suitable for image analysis because the local statistical dependency of an image can be well modeled by Markov properties. MRFs can incorporate *a priori* contextual information or constraints in a quantitative way. The MRF model has been extensively used in various image analysis applications, such as texture synthesis and segmentation, edge detection, and image restoration [69, 70]. We use MRFs to model the dependency of segmented neighboring blocks. As post-processing, MRFs can further improve classification accuracy.

The proposed method is not limited to extracting handwriting from a heterogeneous document. After classification, we can output different contents into different layers. By separating noise, the layer of machine printed text is much cleaner than the original noisy document. Our approach can be used as a document enhancement procedure, which facilitates the further document image analysis tasks, such as zone segmentation and OCR. In this chapter, we demonstrate the effectiveness of our approach on zone segmentation.

The remainder of this chapter is organized as follows. In Section 4.2, we briefly review the previous work on handwriting/machine printed text identification. Noise identification and removal also relates to our work and is reviewed. We present the detailed description of our classification method in Section 4.3. MRF based post-processing is discussed in Section 4.4, and the experimental results are presented in Section 4.5. The effectiveness of our approach for document enhancement is demonstrated in Section 4.6 with the application of zone segmentation. The chapter concludes with a brief summary and a discussion of future work in Section 4.7.

4.2 Related Work

Some work has been accomplished on handwriting/machine printed text identification. The classification is typically performed at the text line [19, 20, 21, 22], word [23], or character level [24, 71]. At the line level, machine printed text lines are typically arranged regularly with a straight baseline, while handwritten text lines are irregular with a varying baseline. Srihari et al. [22] implemented a text line based approach using this characteristic and achieved a classification accuracy of 95%. One advantage of this approach is that it can be used in different scripts (Chinese, English, etc.) with little or no modification. Guo et al. [23] proposed an approach based on the vertical projection profile of the segmented words. They used a hidden Markov model (HMM) as the classifier and achieved a classification accuracy of 97.2%. Although less information is available at the character level, humans can still identify the handwritten and machine printed characters easily, inspiring researchers to pursue classification at the character level. Kuhnke [24] proposed a neural network-based approach with straightness and symmetry as features. Zheng et al. [71] used run-length histogram features to identify handwritten and printed Chinese characters and achieved promising results. In previous work, we implemented a handwriting identification method based on several categories of features and a trained Fisher linear discriminant classifier [25]. However, the problems introduced by noise are not addressed.

Since our approach can be seen as a document enhancement technique, the work on noise removal also relates to our work. Noise may be introduced in document images through (1) physical degradation of the hard-copy documents during creation, and/or storage, and (2) the digitization procedure, such as scanning. If severe enough, either of them can reduce the performance of a document analysis system significantly. Several document degradation models [72, 64, 73], methods for document quality assessment [74, 75], and document enhancement algorithms [76, 77, 78] have been presented in previous work. One common enhancement approach is window-based morphological filtering [76, 77, 78]. Morphological filtering performs a table looking-up procedure to determine an output of ON (black pixel) or OFF (white pixel) for each entry of the table, based on a windowed observation of its neighbors. These algorithms can be further categorized as manually designed, semi-manually designed, or automatically trained approaches. The kFill algorithm, proposed by O'Gorman [78], is a manually designed approach and has been used by several other researchers [74, 79]. Experiments show it is effective for removing salt-and-pepper noise. Liang et al. [80] proposed a semi-manually designed approach with a 3×3 window size. They manually determine some entries to output ON or OFF based on *a priori* observations. The remaining entries are trained to select the optimal output.

It is difficult to manually design a filter with a large window size, and success depends on experience. If both ideal and degraded images are available, optimal filters can be designed by training [77]. After registering the ideal and degraded images at the pixel level, an optimal look-up table can be designed, based on observation of the outputs of each specific windowed context. However, it is difficult to train, store, and retrieve the look-up table when the window size is large. This

63

approach requires both the original and the corresponding degraded images for training. Loce [77] used artificially degraded images generated by models for training, while Kanungo et al. [81, 82, 83] proposed methods for validation and parameter estimation of degradation models. Though the uniformity and sensitivity of their approach has been tested by other researchers [84, 73], no degradation model has been declared to pass the validation. Another problem with morphological approaches is the small window sizes. The most commonly used window size is no larger than 5×5, which is too small to contain enough information for enhancement.

The above approaches only identify and remove small-sized noise components. The removal of large-sized noise components is also addressed in the literature, such as marginal noise removal [85] and show-through removal [86, 87]. It is hard to discriminate noise from compatible sized text. In this work, we treat noise as a distinguished class and use a classification based approach.

4.3   Text Identification

In this section we present our text (machine printed or handwritten) extraction and classification method.

4.3.1   Feature Extraction

Several sets of features are extracted for classification. Table 4.1 lists the descriptions and sizes of the feature sets. Machine printed text, handwriting, and noise have different visual appearances and physical structures. Structural features are

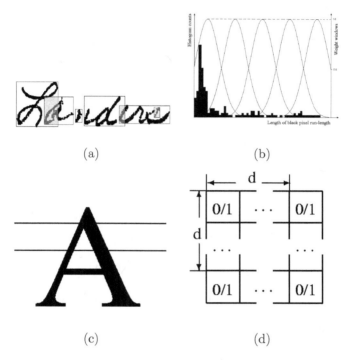

(a) (b)

(c) (d)

Figure 4.1: Illustration of feature extraction. (a) The overlap area of the connected components inside a pattern unit is extracted as a structural feature. (b) Run-length histogram features. (c) Crossing-count features. The crossing counts of the top and bottom horizontal scan lines are 1 and 2, respectively. (d) Bi-level 2 × 2 gram features.

Table 4.1: Features used for machine printed text/handwriting/noise classification

Feature set	Feature description	# of features	# of features selected
Structural	Region size, connected components	18	9
Gabor filter	Stroke orientation	16	4
Run-length histogram	Stroke length	20	5
Crossing-count histogram	Stroke complexity	10	6
Bi-level co-occurrence	Texture	16	2
2×2 gram	Texture	60	5
Total		140	31

extracted to reflect these differences. Gabor filter features and run-length histogram features can capture the difference in stroke orientation and stroke length between handwriting and printed text. Compared with text, noise blocks often have a simple stroke complexity. Therefore, crossing-count histogram features are exploited to model such differences. We aslo take regions of machine printed text, handwriting, and noise blocks as different textures. Two sets of bi-level texture features (bi-level co-occurrence features and bi-level 2×2 gram features) are used for classification. In the following subsections we present these features in detail.

Structural Features

We extract two sets of structural features. The first set includes features related to the physical size of the blocks, such as density of black pixels, width, height, aspect ratio, and area. Suppose the image of the block is $I(x,y)$, $0 \le x < w$, $0 \le y < h$, and w, h are its width and height, respectively. Each pixel in the block has two values: 0 represents background (a white pixel) and 1 represents content (a black pixel). Then the density of the black pixels d is

$$d = \frac{\sum\limits_{x=0}^{w-1} \sum\limits_{y=0}^{h-1} I(x,y)}{w \times h} \qquad (4.1)$$

The sizes of machine printed words are more consistent than those of handwriting and noise on the same page. However, machine printed words on different pages may vary significantly. Therefore, we use a histogram technique to estimate the dominant font size [18], then use the dominant font size to normalize the width (w), height (h), aspect ratio (r), and area (a) of the block.

The second set of structural features is based on the connected components inside the block, such as the mean and variance of the width (m_w and σ_w), height (m_h and σ_h), aspect ratio (m_r and σ_r), and area (m_a and σ_a) of connected components. The sizes of connected components within a machine printed word are more consistent, leading to smaller σ_w and σ_h. For a handwritten word or noise block, the bounding boxes of the connected components tend to overlap with each other, as shown in Fig. 4.1a. For machine printed English words, however, each character forms a connected component not overlapping with others. The overlapping area (the sum of the areas of the gray rectangles in Fig. 4.1a) normalized by the total area of the block is calculated as a feature. We also use the variance of the vertical projection as a feature. In a machine printed text block, the vertical projection profile has obvious valleys and peaks since neighboring characters do not touch each other. However, for a handwritten word or noise block, the vertical projections are much smoother, resulting in smaller variance.

Gabor Filter Features

Gabor filters can represent signals in both the frequency and time domains with minimum uncertainty [88] and have been widely used for texture analysis and segmentation [89]. Researchers found that they match the mammalian visual system very well, which provides further evidence that we can use it in classification tasks. In the spatial and frequency domains, the two-dimensional Gabor filter is defined as

$$g(x,y) = \exp\left\{-\pi\left[\frac{x'^2}{\sigma_x^2} + \frac{y'^2}{\sigma_y^2}\right]\right\} \times \cos\{2\pi(u_0 x + v_0 y)\} \tag{4.2}$$

$$G(u,v) = 2\pi\sigma_x\sigma_y(exp\{-\pi[(u'-u_0')^2\sigma_x^2 + (v'-v_0')^2\sigma_y^2]\} \quad +$$
$$exp\{-\pi[(u'+u_0')^2\sigma_x^2 + (v'+v_0')^2\sigma_y^2]\}) \tag{4.3}$$

where $x' = -x\sin\theta + y\cos\theta$, $y' = -x\cos\theta - y\sin\theta$, $u' = u\sin\theta - v\cos\theta$, $v' = -u\cos\theta - v\sin\theta$, $u_0' = -u_0\sin\theta + v_0\cos\theta$, $v_0' = -u_0\cos\theta - v_0\sin\theta$, $u_0 = f\cos\theta$, and $v_0 = f\sin\theta$. Here f and θ are two parameters, representing the central frequency and orientation of the Gabor filter.

The variances of the filtered images are taken as features. In our experiments 16 Gabor filters with different orientations $\theta_k = k \times 180/N$, $k = 1, 2, \ldots 16$, are used, which generate 16 features.

Run-length Histogram Features

Run-length histogram features are proposed in [25] for machine printed/handwritten Chinese character classification. These features are used in our case to capture the difference between the stroke lengths of machine printed text, handwriting, and noise blocks. First, black pixel run-lengths in four directions, including horizontal, vertical, major diagonal, and minor diagonal, are extracted. We then calculate four histograms of run-lengths for these four directions, as shown in Fig. 4.1b. To get scale-invariant features, we normalize the histograms. Suppose C_k, $k = 1, 2, ..., N$, is the number of runs with length k, and N is the maximal length of all possible runs, then the normalized histogram C'_k is

$$C'_k = \frac{C_k}{\sum\limits_{i=1}^{N} C_i} \tag{4.4}$$

We then divide the histogram into five bins with equal width and use five Gaussian-shaped weight windows to get the final features (Fig. 4.1b). Taking the horizontal run-length histogram as an example, the run-length histogram feature Rh_i is calculated as

$$Rh_i = \sum_{k=1}^{w} G(k; u_i, \sigma)C'_k, \quad i = 1, 2, 3, 4, 5 \tag{4.5}$$

where w is the width of the block (the maximal length of all possible horizontal run-lengths) and $G(k; u_i, \sigma)$ is a Gaussian-shaped function:

$$G(k; u_i, \sigma) = exp\left\{-\frac{(k - u_i)^2}{2\sigma^2}\right\} \tag{4.6}$$

As shown in Fig. 4.1b, σ is chosen so the weight on each bin border is 0.5. Another alternative is to use rectangular windows without overlap between neighboring bins. Experiments show that the extracted features with Gaussian weighted windows are more robust. Five features are extracted in each direction, leading to 20 features.

Crossing-Count Histogram Features

A crossing count is the number of times the pixel value changes from 0 (white pixel) to 1 (black pixel) along a horizontal or vertical raster scan line. As shown in Fig. 4.1c, the crossing counts of the top and bottom horizontal scan lines are 1 and 2, respectively. Crossing counts can be used to measure stroke complexity [90, 25]. In our approach, first the crossing count for each horizontal and vertical scan line is calculated. Similarly, we get two histograms for the horizontal and vertical crossing counts respectively. The same technique (as in extracting the run-length histogram features) is exploited to get the final features from the histograms. A total of 10 features are extracted.

Bi-level Co-occurrence Features

A co-occurrence count is the number of times a given pair of pixels occurs at a fixed distance and orientation [91]. In the case of binary images, the possible co-occurrence pairs are white-white, black-white, white-black, and black-black. In our case, we are concerned primarily with the foreground. Since the white background region often accounts for up to 80% of a document page, the occurrence frequency of white-white or white-black pixel pairs will always be much higher than that of black-black pairs. The black-black pairs carry most of the information. To eliminate the redundancy and reduce the effects of over-emphasizing the background, we consider only black-black pairs. Four different orientations (horizontal, vertical, major diagonal and minor diagonal) and four distance levels (1, 2, 4, and 8 pixels)

70

are used to classify (16 features total). The horizontal co-occurrence count $C_h(d)$, for example, is defined as

$$C_h(d) = \sum_x \sum_y I(x,y)I(x+d,y), d = 1, 2, 4, 8 \tag{4.7}$$

$I(x,y) = 0$ for white pixels; therefore only black-black pixel pairs contribute. For a fixed distance d, we normalize the occurrence by dividing by the sum of the occurrences in all four directions.

Bi-level 2×2 gram Features

The N×M grams were first introduced in the context of image classification and retrieval [92]. An N×M gram extends the one-dimensional co-occurrence feature to the two-dimensional case. We only consider 2×2 grams, which count the numbers of occurrences of the patterns shown in Fig. 4.1d. The cells labeled 0/1 should take specific values, and the values of other cells are irrelevant. Therefore, there are $2^4 = 16$ patterns for each distance d. Like the co-occurrence features, the all white patterns are removed to reduce over-emphasis on the background. For a fixed distance, the occurrences are normalized by dividing by the sum of all occurrences. Four distances (1, 2, 4, and 8 pixels) are chosen, generating $4 \times 15 = 60$ features.

4.3.2 Feature Selection

There are two purposes for feature selection. The first involves reducing the computation needed for feature extraction and classification. As shown in Table 4.1, we extract a total of 140 features from the segmented blocks. Though these features are

71

designed to distinguish between different types of blocks, some features may contain more information. Using only a small set of the most powerful features reduces the time for feature extraction and classification. The second purpose is to alleviate the curse of dimensionality problem. When the number of training samples is limited, using a large feature set may decrease the generality of a classifier [93]. The larger the feature set, the more training samples are needed. Therefore, we perform feature selection before feeding the features to the classifier.

We use a forward search algorithm to perform feature selection [94]. We first divide the whole feature set $\mathcal{F}$ into a currently selected feature set $\mathcal{F}_s$ and an un-selected feature set $\mathcal{F}_n$ which satisfy

$$\mathcal{F}_s \cap \mathcal{F}_n = \Phi \tag{4.8}$$

$$\mathcal{F}_s \cup \mathcal{F}_n = \mathcal{F} \tag{4.9}$$

The selection procedure can then be described as

1. Set $\mathcal{F}_s = \Phi$, and $\mathcal{F}_n = \mathcal{F}$.

2. Label all features in $\mathcal{F}_n$ as un-tested.

3. Select one un-tested feature $f \in \mathcal{F}_n$ and label it as tested.

4. Put f and $\mathcal{F}_s$ together and generate a temporary selected feature set $\mathcal{F}_s^f$.

5. Estimate the classification accuracy with feature set $\mathcal{F}_s^f$ using a 1-NN classifier and leave-one-out cross validation technique. Basically, at each iteration only one sample is used for testing, while the others are used for training. We

repeat this process until all samples have been used as testing samples once. The average accuracy for all iterations is taken as the estimated accuracy for the current feature set. The leave-one-out cross validation technique can estimate the accuracy of a classifier with small variation [93].

6. If there are un-tested features in $\mathcal{F}_n$, go to step 3.

7. Find a feature $\hat{f} \in \mathcal{F}_n$, such that the corresponding temporary feature set $\mathcal{F}_s^f$ has the highest classification accuracy:

$$\hat{f} = \arg \max_{f \in \mathcal{F}_n} Accuracy(\mathcal{F}_s^f) \qquad (4.10)$$

then move $\hat{f}$ from $\mathcal{F}_n$ to $\mathcal{F}_s$.

8. If $\mathcal{F}_n \neq \Phi$, go to step 2; otherwise exit.

We use LNKnet pattern classification software to conduct our feature selection experiments [95]. LNKnet provides several classifiers, such as likelihood classifiers, k-NN classifiers, and neural network classifiers, and several feature selection algorithms such as forward search, backward search, and forward and backward search. Feature selection can be an extremely expensive task. Considering the large number of feature sets to evaluate, and the number of classifiers to train, the lightweight forward feature selection algorithm and 1-NN classifier, which does not need training, are used in our feature selection experiment.

We collected about 1,500 blocks for each class. As shown in Fig. 4.2a, when the number of selected features increases, the error rate decreases sharply at first. The trend reverses at some point. The best classification is achieved when only 31

73

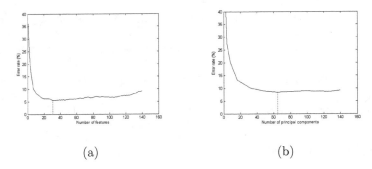

(a)   (b)

Figure 4.2: Feature analysis. (a) Feature selection: the best classification result is achieved when 31 features are selected. (b) PCA: the best classification result is achieved when 64 principal components are used.

features are selected, with an error rate of 5.7%. When all features are used, the error rate increases to 9.2% due to the limited number of training samples and large feature set. The last column in Table 4.1 lists the number of features selected in each set. It shows that texture features, such as bi-level co-occurrence and 2×2 grams, are less discriminating than other feature sets, mainly because of the small region size. Only 1/8 of the bi-level co-occurrence features and 1/12 of the 2×2 gram features are selected. Crossing-count histogram features and structural features are more effective, with more than half of the original features in both sets selected in the final feature set.

Principal component analysis (PCA) is another technique for reducing feature dimension [93]. To extract the first n principal components, we need to search a subspace of dimension n with basis $\mathbf{w}$. Suppose the mean is already removed from the feature vector $\mathbf{X}$, and let the projection of $\mathbf{X}$ onto this subspace be $\hat{\mathbf{X}}$

$$\hat{\mathbf{X}} = (\mathbf{w}_1^T \mathbf{X})\mathbf{w}_1 + (\mathbf{w}_2^T \mathbf{X})\mathbf{w}_2 + \ldots + (\mathbf{w}_n^T \mathbf{X})\mathbf{w}_n \qquad (4.11)$$

PCA finds the optimal subspace $\hat{\mathbf{w}}$ such that the energy contained in $\hat{\mathbf{X}}$ is maximized:

$$\hat{\mathbf{w}} = \arg \max_{\mathbf{w}_1, \ldots, \mathbf{w}_n} \sum_{i=1}^{n} Var \left[\hat{\mathbf{X}}_i \right]$$

$$\text{s.t. } \mathbf{w}_i^T \mathbf{w}_j = \begin{cases} 1 & \text{if } i = j \\ 0 & \text{if } i \neq j \end{cases} \tag{4.12}$$

The optimal basis is the first n eigenvectors of the covariance matrix of $\mathbf{X}$, corresponding to the first n eigenvalues [93]. The first n principal components are $P_i = \mathbf{w}_i^T \mathbf{X}, i = 1, \ldots, n$. The idea of PCA is to concentrate the energy into the first several principal components. Assuming the classification information is contained in the energy, the first several principal components are more powerful than the remaining components. Furthermore, PCA analysis can remove the correlation among features.

As in the feature selection experiment, we use the 1-NN classifier and the leave-one-out technique to estimate the classification accuracy. Fig. 4.2b shows the classification error rate versus the number of principal components used. As in feature selection, the error rate reduces quickly at first until 16 principal components added. The minimal error rate, 8.5%, is achieved when 64 principal components are used. Compared with the minimum error rate of 5.7% achieved by the feature selection technique, PCA is not as powerful as feature selection in this problem. Furthermore, to perform PCA, all features must be extracted first. However, for feature selection, we only need to extract the desired features, which would increase the feature extraction speed. Therefore, in the following, we do classification on the 31 selected features.

4.3.3 Classification

Compared with the neural network (NN) and the support vector machine (SVM), the Fisher linear discriminant classifier is easier to train, faster to classify, needs fewer training samples, and does not suffer from the over-training problems. According to the comparison experiment in Subsection 4.5.2, the SVM classifier performs slightly better than the Fisher linear discriminant classifier, but the latter is much faster. We therefore use it for classification. For a feature vector $\mathbf{X}$, the Fisher linear discriminant classifier projects $\mathbf{X}$ onto one dimension Y in direction $\mathbf{W}$

$$Y = \mathbf{W}^T \mathbf{X} \tag{4.13}$$

The Fisher criterion finds the optimal projection direction $\mathbf{W}_o$ by maximizing the ratio of the between-class scatter to the within-class scatter, which benefits the classification. Let $\mathbf{S}_w$ and $\mathbf{S}_b$ be the within- and between-class scatter matrices respectively,

$$\mathbf{S}_w = \sum_{k=1}^{K} \sum_{\mathbf{x} \in \text{class } k} (\mathbf{x} - \mathbf{u}_k)(\mathbf{x} - \mathbf{u}_k)^T \tag{4.14}$$

$$\mathbf{S}_b = \sum_{k=1}^{K} (\mathbf{u}_k - \mathbf{u}_0)(\mathbf{u}_k - \mathbf{u}_0)^T \tag{4.15}$$

$$\mathbf{u}_0 = \frac{1}{K} \sum_{k=1}^{K} \mathbf{u}_k \tag{4.16}$$

where $\mathbf{u}_k$ is the mean vector of the kth class, $\mathbf{u}_0$ is the global mean vector, and K is the number of classes. The optimal projection direction is the eigenvector of $\mathbf{S}_w^{-1}\mathbf{S}_b$ corresponding to its largest eigenvalue [93]. For a two-class classification problem, we do not need to calculate the eigenvectors of $\mathbf{S}_w^{-1}\mathbf{S}_b$. It is shown that the optimal projection direction is

$$\mathbf{W}_o = \mathbf{S}_w^{-1}(\mathbf{u}_1 - \mathbf{u}_2)$$ (4.17)

Let Y_1 and Y_2 be the projections of two classes and let $E[Y_1]$ and $E[Y_2]$ be the means of Y_1 and Y_2 respectively. Suppose $E[Y_1] > E[Y_2]$, then the decision can be made as

$$C(\mathbf{X}) = \begin{cases} \text{class 1} & \text{If } Y > (E[Y_1] + E[Y_2])/2 \\ \text{class 2} & \text{Otherwise} \end{cases}$$ (4.18)

It is known that if the feature vector $\mathbf{X}$ is jointly Gaussian distributed, and the two classes have the same covariance matrices, then the Fisher linear discriminant classifier is optimal in a minimum classification error sense [93].

The Fisher linear discriminant classifier is often used for two-class classification problems. Although it can be extended to multi-class classification (three classes here), the classification accuracy decreases due to the overlap between neighboring classes. Therefore, we use three Fisher linear discriminant classifiers, each optimized for a two-class classification problem (machine printed text/handwriting, machine printed text/noise, and handwriting/noise). Each classifier outputs a classification confidence, and the final decision is made by fusing the outputs of all three classifiers.

4.3.4 Classification Confidence

In a Fisher linear discriminant classifier, the feature vector is projected onto an axis on which the ratio of between-class scatter to within-class scatter is maximized. According to the central limit theorem [52], the distribution of the projection can be approximated by a Gaussian distribution, if no feature has dominant variance over the others,

$$f_Y(y) = \frac{1}{\sqrt{2\pi}\sigma} \exp\left[-\frac{1}{2}\left(\frac{y-m}{\sigma}\right)^2\right],$$ (4.19)

where $f_Y(y)$ is the probability density function of the projection. The parameters m and σ can be estimated from training samples. The classification confidence $C_{i,j}$ of class i using classifier j is defined as

$$C_{i,j} = \begin{cases} \frac{f_Y(y|\mathbf{x}\in\text{class } i)}{f_Y(y|\mathbf{x}\in\text{class } i)+f_Y(y|\mathbf{x}\in\text{another class})} & \text{If } i \text{ is applicable for classifier } j. \\ 0 & \text{Otherwise} \end{cases}$$ (4.20)

where i is the class label and j represents the trained classifiers. If a classifier is trained to classes 1 and 2, its output is not applicable to estimating the classification confidence of class 3. Therefore, $C_{3,j} = 0$. The final classification confidence is defined as

$$C_i = \frac{1}{2}\sum_{j=1}^{3} C_{i,j}$$ (4.21)

$C_{i,j} \in [0,1]$ for the two applicable classifiers and $C_{i,j} = 0$ for the third classifier, $C_i \in [0,1]$. However, C_i is not a good estimate of the *a posteriori* probability since $\sum_{i=1}^{3} C_i = 1.5$ instead of 1. We can take C_i as an estimate of a non-decreasing function of the *a posteriori* probability, which is a kind of generalized classification confidence [96].

Fig. 4.3 shows the word segmentation and classification results (with the Fisher linear discriminant classifier) for the whole and parts of a document image, with blue, red, and green representing machine printed text, handwriting, and noise respectively. We can see that most of the blocks are correctly classified. However some blocks are misclassified due to an overlap in the feature space. For example,

78

some noise blocks are classified as handwriting in Fig. 4.3b, and some small printed words are classified as noise in Fig. 4.3c. Since little information is available in small areas, it is difficult to get good results. In the next section, we present a method of Markov random field (MRF) based post-processing to refine the classification by incorporating contextual information.

4.4 MRF-Based Post-Processing

4.4.1 Background

Let $\mathbf{X}$ denote the random field defined on Ω, and let Γ denote the set of all possible configurations of $\mathbf{X}$ on Ω. $\mathbf{X}$ is an MRF with respect to the neighborhood η if it has the following Markov property

$$\Pr(\mathbf{X} = \mathbf{x}) > 0 \qquad \text{for all} \quad \mathbf{x} \in \Gamma \tag{4.22}$$

$$P(x_s|x_r, r \in \Omega, r \neq s) = P(x_s|x_r, r \in \eta) \tag{4.23}$$

Compared with Markov chains, one difficulty with MRFs is that they have no chain rule. The joint probability $P(\mathbf{X} = \mathbf{x})$ cannot be recursively written in terms of local conditional probabilities $P(x_s|x_r, r \in \eta)$. Therefore, it is difficult to get an optimal estimate of the MRF $\hat{\mathbf{X}}$ which maximizes the *a posteriori* probability

$$\hat{\mathbf{X}} = \arg \max_{\mathbf{X}} P(\mathbf{X}|\mathbf{Y}) \tag{4.24}$$

The establishment of the connection between the MRF and Gibbs distribution provides a way to optimize the MRF. To maximize the *a posteriori* probability of the

(a)

Gene Russell was with me when I calls
The gist of what I said was as follow
media by CEOs do make sense in the pr
need to select the ground carefully i
appropriate advisers. I suggested th
the Waxman-Hatch-Packwood legislation
advertising because it presents first
are so dear to the hearts of publishe
that the footing would not be as good
ground, heart disease and the Surgeon

(b)

February 14, 1983

Industry Contacts With Editors

to reports on my call of February 11 to
ng the attached letters. Both letters
y s and both deal with contacts by the
of publications in which cigarette ads

the letters is from Curt Judge to Sam C
shown to the TI Executive Committee. I
nce with respect to a "rebuttal" which
to editors of Time Magazine on the sub

(c)

Figure 4.3: Word block segmentation and classification results, with blue, red, and green representing machine printed text, handwriting, and noise respectively. (a) A whole document image. (b) and (c) Two parts of the image in (a).

MRF, we need to minimize the total energy of the corresponding Gibbs distribution

$$\hat{\mathbf{X}} = \arg\min_{\mathbf{X}} \sum_{c \in \mathcal{C}} V_c(\mathbf{X}) \tag{4.25}$$

Here, a *clique c* is defined as a subset of sites in which every pair of distinct sites are neighbors. The *clique potential* $V_c(\mathbf{X})$ is the energy associated with a clique and depends on the local configuration of clique c. Therefore, the optimization problem (4.24) is converted to another optimization problem (4.25). The information about the observation $\mathbf{Y}$ is contained in the clique system.

In the study of MRFs, the problems are often posed as labeling problems in which a set of labels are assigned to sites of an MRF [70]. In our problem, each block constitutes a site of an MRF. A label (machine printed text, handwriting, or noise) is assigned to each block, and context information (encoded by the MRF model) is used to flip the labels so that the total energy of the corresponding Gibbs distribution is minimized. Relaxation algorithms are often used for MRF optimization [70].

4.4.2 Clique Definition

As shown in (4.25), the MRF is totally determined by clique c and clique potential $V_c(\mathbf{X})$. The design of the clique and its potential is crucial, but a systematic method is not yet available. In our case, machine printed text, handwriting, and noise exhibit different patterns of geometric relationships. Our definition of cliques reflects these differences.

Printed words often form horizontal (or vertical) text lines. Clique C_p is defined in Fig. 4.4a, which models contextual constraints on neighboring machine printed words. We first define the *connection* between word blocks i and j. As shown in Fig. 4.4a, O_v is the vertical overlap between two blocks, and D_h is the horizontal distance between two blocks. The distance between block i and j is

$$D(i,j) = |D_h(i,j) - G_w| + |H_i - H_j| + |Ch_i - Ch_j| \tag{4.26}$$

where $D_h(i,j)$ is the horizontal distances between words i and j, G_w is the estimated average word gap in the whole document, H_i and H_j are the heights of blocks i and j respectively, and Ch_i and Ch_j are the vertical centers of the two blocks. Two blocks are connected if they satisfy

1. $O_v \geq min(H_i, H_j)/2$

2. $0 \leq D_h \leq 2G_w$

3. $D(i,j) < T_p$, where T_p is a threshold, which is not sensitive to post-processing.

After defining the connection between two blocks we can construct a graph in which nodes represent blocks and edges link two connected nodes. If a node is connected with more than one node on one side (left or right), we keep only the edge with the smallest distance. Clique C_p can be represented by nodes together with their left and right neighbors. If we cannot find neighbors on the left or/and right sides, the corresponding neighbor is set to NULL.

Noise blocks exhibit random patterns in geometric relationships and tend to overlap or in close proximity. As shown in Fig. 4.4b, the noise block labeled "Center" is overlapped with blocks 1, 2, 3, and is close to block 4. Clique C_n is defined

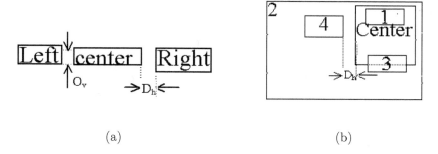

(a)   (b)

Figure 4.4: Clique definition. (a) C_p for horizontally arranged machine printed words. (b) C_n for noise blocks.

primarily for noise blocks. Similarly, the distance between two blocks is defined as

$$D(i,j) = \max(D_h(i,j), D_v(i,j)) \tag{4.27}$$

where $D_h(i,j) = \max(L_i, L_j) - \min(R_i, R_j)$, $D_v(i,j) = \max(T_i, T_j) - \min(B_i, B_j)$, and L, R, T, B are the left, right, top, and bottom coordinates of the corresponding blocks. If two blocks overlap in the horizontal or vertical direction, then $D_h(i,j) < 0$ or $D_v(i,j) < 0$. Blocks i and j are connected if, and only if, $D(i,j) < T_n$, where T_n is a threshold. If T_n is too big, incorrect label flipping of noise and handwriting between two printed text lines may happen. If T_n is too small, the contextual constraints on the noise blocks cannot be used fully. We set T_n as half of the dominant character height (about 10 pixels in our experiments). Each node, together with all nodes connected to it, defines clique C_n. The number of connected nodes may vary from 0 to almost 10, depending on the size of the block. As an approximation, we consider only the first four nearest connected neighbors. If the number of neighbors is less than four, we set the corresponding neighbors to NULL.

The geometric constraint on handwriting has weaker horizontal or vertical

83

structure than machine printed words, thus it is partially reflected in both cliques C_p and C_n. Therefore, we do not define a specific clique for handwriting.

4.4.3 Clique Potential

Clique potential is the energy associated with a clique. We assign high energy to an undesirable configuration of the clique and low energy to a preferred configuration. For example, an undesired configuration of clique C_p (as shown in Fig. 4.4a) is that the left and right blocks are labeled as printed text and the center block as noise. Flipping the label of the center block from noise to printed text would achieve a more preferred configuration, and reduce the total energy. Another undesirable configuration occurs when all blocks are labeled as printed text for the clique C_n in Fig. 4.4b. It should have higher energy than the configuration in which all blocks are labeled as noise. In many applications the clique potentials are defined ad hoc. One systematic way is to define clique potential as the occurrence frequency of each clique in the training set, which can be expressed as a function of local conditional probabilities. Based on this idea, we define two clique potentials $V_p(c)$ and $V_n(c)$ for cliques C_p and C_n as

$$V_p(c) = -\frac{P(X_l, X_c, X_r)}{(P(X_l)P(X_c)P(X_r))^w}, \tag{4.28}$$

$$V_n(c) = -\frac{P(X_c, X_1, X_2, X_3, X_4)}{(P(X_c)P(X_1)P(X_2)P(X_3)P(X_4))^w}, \tag{4.29}$$

where X_l, X_c and X_r are labels for the left, center, and right blocks of clique c, w is a constant, and X_i, $i = 1, 2, 3, 4$, is the label of the ith nearest block. The energy of the corresponding Gibbs distribution is

$$U(\mathbf{X}|\mathbf{Y}) = w_s \sum_{s \in \Omega} [-P(x_s|y_s)] + w_p \sum_{c \in C_p} V_p(c) + w_n \sum_{c \in C_n} V_n(c), \qquad (4.30)$$

where w_s, w_p, and w_n are weights which adjust the relative importance of classification confidence and contextual information for cliques C_p and C_n. If $w_s = 1$,
$w_p = 0$, and $w_n = 0$, no contextual information is used; with increase in w_p and w_n,
more contextual information is emphasized. If we set $w_p = w_n = \infty$, or equivalently
$w_s = 0$, no classification confidence is used.

In the following experiments, we want to use MRFs for word block labeling. The number of handwritten words is much smaller than the other two types,
leading to a lower estimated frequency of cliques with handwriting. As a result,
the optimization tends to label handwritten words as machine printed text or
noise. Therefore, we regularize the estimated clique frequency $P(X_l, X_c, X_r)$ and
$P(X_c, X_1, X_2, X_3, X_4)$ by dividing by the product of the probabilities of the word
block labels which comprise the clique. The above regularization is similar to the
previous approach [97], where w is set to 1. In our case, w is changeable; increasing
w will emphasize handwritten words. Our clique potential definition is systematic
and can be optimized for different applications.

After defining the cliques and the corresponding clique potential, we can search
the optimal configuration of the labels of all blocks, so the total energy of the corresponding Gibbs distribution is minimized. Relaxation algorithms are often used for
MRF optimization. There are two types of relaxation algorithms: stochastic and
deterministic [70]. Stochastic algorithms can always converge to the global optimal

solution if some constraints are satisfied. They are, however, computationally demanding. Deterministic algorithms are simpler, but only converge to local optimal solutions depending on the initial value. In our experiments, highest confidence first (HCF), a deterministic approach, is used for MRF optimization due to its fast speed and good performance [98]. In each iteration of the HCF algorithm, only one block is chosen to flip its label such that the total energy reduces the largest. It repeats this procedure until no single flipping can further reduce the total energy. Since each flipping would reduce the energy and the energy is bounded below, the HCF algorithm converges in a finite number of steps. Fig. 4.5 is an example of the refined classification results after post-processing. Compared with Fig. 4.3, we can see in Figs. 4.5a and (b) that most misclassified noise blocks are corrected, with a few exceptions due to their having fewer constraints. The misclassified small machine printed words are all corrected in Fig. 4.5c.

4.5 Experiments

4.5.1 Data Set

We collected a total of 318 business letters from the tobacco industry litigation archives. These document images are noisy with a significant amount of handwritten annotations and signatures, a few logos, and no figures or tables. Currently, we identify three classes: machine printed text, handwriting, and noise. We used 224 images for training and the remaining 94 for testing. There are about 1,500 handwritten words in the training set. Since much more machine printed and noise

(a)

Gene Russell was with me when I calle
The gist of what I said was as follow
media by CEOs do make sense in the pr
need to select the ground carefully a
appropriate advisers. I suggested th
the Waxman-Hatch-Packwood legislation
advertising because it presents first
are so dear to the hearts of publishe
that the footing would not be as good
ground, heart disease and the Surgeon

(b)

February 14, 1983

Industry Contacts With Editors

to reports on my call of February 11 to
ig the attached letters. Both letters
a and both deal with contacts by the
of publications in which cigarette ads

ne letters is from Curt Judge to Sam C
nown to the TI Executive Committee. I
ice with respect to a "rebuttal" which
to editors of Time Magazine on the sub

(c)

Figure 4.5: Word block classification results after post-processing with blue, red, and green representing machine printed text, handwriting, and noise respectively. (a) The whole document image. (b) and (c) Two parts of the image in (a).

blocks are present, we randomly selected about the same number of blocks of each type for training. We used *accuracy* and *precision* as metrics to evaluate the result:

$$\text{Accuracy of type i} = \frac{\text{\# of correctly classified blocks of type i}}{\text{\# of blocks of type i}}, \tag{4.31}$$

$$\text{Precision of type i} = \frac{\text{\# of correctly classified blocks of type i}}{\text{\# of blocks classified as type i}}. \tag{4.32}$$

4.5.2 Classifier Comparison

In this section, we compare the performance of three different classifiers: the k-NN classifier, the Fisher linear discriminant classifier, and the SVM classifier. The SVM classifier is based on VC dimension theory and structural risk minimization theory of statistical learning [99]. A public domain SVM tool, LibSVM, is used in the following experiment [100]. The N-fold ($N = 10$ in our experiment) verification technique, a variation of the leave-one-out technique, is used to estimate the classification accuracy. Instead of holding one sample for testing at each iteration, it first divides the data set into N groups ($N = 10$ in our experiment), then holds one group of samples for testing and the remaining groups for training. Table 4.2 shows the classification accuracies of all the classifiers. We can see that the SVM classifier achieved the highest accuracy. Considering the large variance, the improvement is not significant. The variance of the classification accuracy of all classifiers is the smallest for printed text and the largest for handwriting, indicating that the printed text is more compact in the feature space. Among all three classifiers, the Fisher linear discriminant classifier is the fastest since it needs only one vector multiplication to perform a classification. Therefore, we use the Fisher linear discriminant

Table 4.2: Performance comparison of three different classifiers: the k-NN classifier, the Fisher linear discriminant classifier, and the SVM classifier.

	# of blocks	the k-NN classifier			the Fisher classifier			the SVM classifier		
		Correct	Accuracy	Var	Correct	Accuracy	Variance	Correct	Accuracy	Var
Printed text	1,519	1,489	98.0%	1.4%	1,473	97.0%	1.1%	1,480	97.4%	1.2%
Handwriting	1,518	1,390	91.6%	2.3%	1,410	92.9%	2.2%	1,435	94.5%	2.1%
Noise	1,512	1,406	93.0%	2.0%	1,451	96.0%	1.5%	1,453	96.1%	1.2%
Overall	4,549	4,285	94.2%	1.3%	4,344	95.5%	0.9%	4,368	96.0%	0.9%

Table 4.3: Single word block classification.

	# of blocks	Percentage	# of correctly classified blocks	# of misclassified blocks	Accuracy	Precision
Printed text	19,227	66.9%	18,446	781	95.9%	99.5%
Handwriting	701	2.4%	653	48	93.2%	62.9%
Noise	8,802	30.7%	8,522	280	96.8%	93.0%
Overall	28,730	100.0%	27,621	1,109	96.1%	N/A

classifier for the rest of the experiments.

The classification result on the test set of 94 images, using the Fisher linear discriminant classifier, is shown in Table 4.3. The accuracies on all three classes range from 93.2% to 96.8%, with an overall accuracy of 96.1%. While this overall accuracy is high, we notice that the precision for handwriting is low (62.9%). This is mainly because the number of handwritten words in the testing set is small. Even small percentages of mis-classification of machine printed text and noise as handwriting will significantly decrease the precision of handwriting.

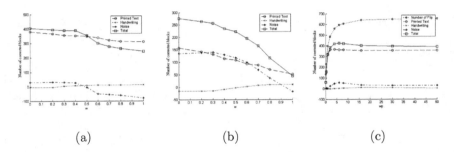

(a)   (b)   (c)

Figure 4.6: MRF-based post-processing. (a) Number of corrected blocks using clique C_p. (b) Number of corrected blocks using clique C_n. (c) Number of corrected blocks using clique C_p and classification confidence.

4.5.3   Post-processing Using MRFs

In the following experiments we investigate how MRFs can improve classification accuracy. In the first run, we set $w_s = 0$, $w_n = 0$ and $w_p = 1$ to show the effectiveness of clique C_p. Fig. 4.6a shows the number of corrected blocks, which were previously misclassified, with change in w. As expected, C_p is effective for machine printed words, but not as effective for handwriting and noise. When $w = 0.3$ (under this condition, the classification accuracy of all three classes increases), 355 (46%) of the previously misclassified machine printed words are corrected. When w increases, handwriting is emphasized more, leading to higher classification accuracy of handwriting, and lower accuracy of machine printed words and noise. In practice, w can be adjusted to optimize the overall accuracy.

In the second run, we test the effectiveness of clique C_n by setting $w_s = 0$, $w_p = 0$, and $w_n = 1$. As shown in Fig. 4.6b, clique C_n effectively corrects classification errors of noise blocks. The classification error of noise blocks is greatly reduced

90

when w is small. For $w = 0.6$ (under this condition, the classification accuracy of all classes increases), the number of misclassified noise blocks is reduced by 99 (35%). C_n can also correct some classification errors of machine printed words, but is less effective than C_p as shown in Fig. 4.6a.

The third run tests the effectiveness of classification confidence for post-processing. Fig. 4.6c shows post-processing results by adjusting w_p when $w = 0.3$, $w_n = 0$, and $w_s = 1$. Adjusting w_p will change the total flip number. When $w_p = 0$, the energy reaches the minimum with the initial labels, and the total flip number is 0. When w_p increases, more emphasis is placed on contextual information, and the flip number increases. When $w_p \rightarrow +\infty$, it converges to the case of $w_p = 1$ and $w_s = 0$, the setting of the first run. The maximal overall classification accuracy is achieved when $w_p = 6$. Compared with the first run, the total number of corrected blocks increases from 389 to 424 by incorporating classification confidence. Similar results are achieved by combining classification confidence with clique C_n.

In the last run, we fix $w_s = 1$ and manually adjust w, w_p, and w_n to optimize the overall classification accuracy. The final parameters we chose are $w = 0.39$, $w_p = 5$, and $w_n = 4$. Table 4.4 shows the results after post-processing. The "Error Reduction Rate" in Table 4.4 is defined as follows:

$$
\begin{matrix} \text{Error} \\ \text{Reduction} \\ \text{Rate} \end{matrix} = \frac{\text{\# of Errors Before Post-Processing} \ - \ \text{\# of Errors After Post-Processing}}{\text{\# of Error Before Post-Processing}}
$$

$$(4.33)$$

The error rate reduces to about half of the original for both machine printed text and noise, but increases slightly for handwriting. However, compared with Ta-

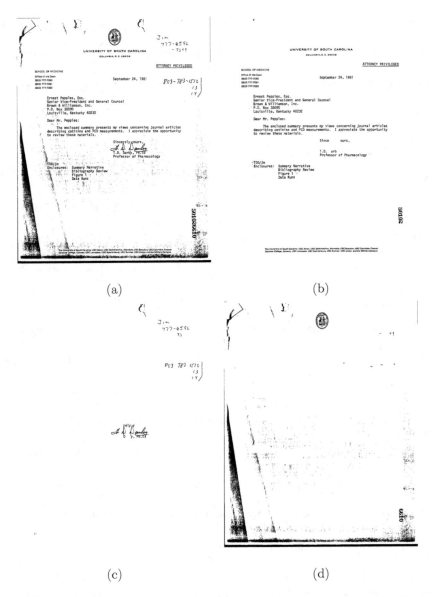

Figure 4.7: An example of machine printed text and handwriting identification from noisy documents. (a) The original document image. (b) Machine printed text. (c) Handwriting. (d) Noise. The logo is classified as noise since currently we only consider three classes.

Table 4.4: Word block classification after MRF based post-processing.

	# of blocks	# of correctly classified blocks	# of misclassified blocks	Reduction of misclassified blocks	Error reduction rate	Accuracy	Precision
Printed text	19,227	18,835	392	389	49.8%	98.0%	99.7%
Handwriting	701	652	49	-1	-2.1%	93.0%	83.3%
Noise	8,802	8,682	120	160	57.1%	98.6%	96.0%
Total	28,730	28,169	561	548	49.4%	98.1%	N/A

ble 4.3, the precision of handwriting increases from 62.9% to 83.3% due to fewer machine printed text and noise mis-classifications as handwriting. The overall accuracy increases from 96.1% to 98.1%.

Fig. 4.7 shows an example of machine printed text and handwriting identification from noisy documents. To display the classification results clearly, we decompose the classified image into three layers, representing machine printed text (Fig. 4.7b), handwriting (Fig. 4.7c), and noise (Fig. 4.7d) respectively. The result is good with few mis-classifications.

Our approach is general and can be extended to other languages with minor modification. Fig. 4.8 shows identification results for a Chinese document. In Chinese, there is no clear definition of words and no spaces between neighboring words. Therefore, the parameters of our word segmentation module are adjusted to get characters. We only need to retrain the classifiers; the post-processing module is intact. We can see that most handwriting and noise blocks are classified correctly, but several machine printed digits are misclassified as handwriting. On the right margin of the document, some machine printed text is identified as noise due to touching.

Our approach is fast; the averaging processing time for a business letter scanned at 300 dpi is about 2-3 seconds on a PC with 1.7 GHz CPU and 1.0 GB memory.

4.6 Application to Zone Segmentation in Noisy Images

The proposed method is not limited to extract handwriting from a heterogeneous document. After classification, we can output different contents into different layers. By separating noise, the layer of machine printed text becomes cleaner than the original noisy document. Therefore, our approach can be used as a document enhancement procedure, which facilitates further document image analysis tasks, such as zone segmentation and OCR.

In this section, we show that our method can improve general zone segmentation results after removing identified noise. We evaluated two widely used zone segmentation algorithms: the Docstrum algorithm [18] and ScanSoft SDK, a commercial OCR software package [101]. Many different zone segmentation evaluation metrics have been proposed in previous work. Kanai et al. [102] evaluated zone segmentation accuracy from the OCR aspect. Any zone splitting and merging, if it does not affect the reading order of the text, is not penalized. The approach of Mao and Kanungo [103] is based on text lines, which penalizes only horizontal text line splitting and merging, since it will change the reading order of text. Randriamasy et al. [104] proposed an evaluation method based on multiple ground truth, which

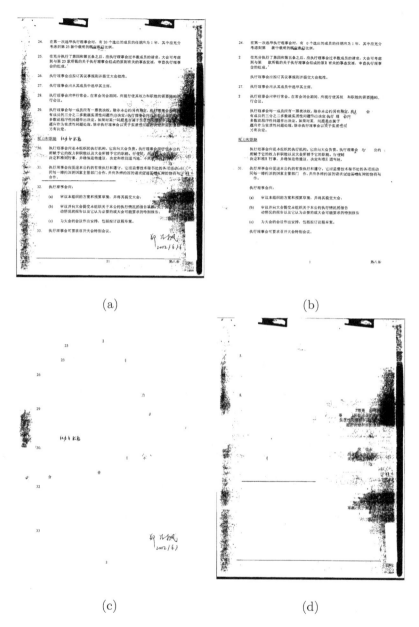

Figure 4.8: An example of machine printed text and handwriting identification from Chinese documents. (a) Original Chinese document image. (b) Machine printed text. (c) Handwriting. (d) Noise.

is very expensive. Liang's approach is performed at the zone level [76]. After finding the correspondence between the segmented and ground-truthed zones, any large enough difference is penalized. We use Liang's scheme in our experiment since we focus on zone segmentation. From the OCR perspective, vertical splitting or merging of different zones should not be penalized even when these zones have different physical and semantic properties. From the point of view for zone segmentation, it should be penalized.

There are 1,374 machine printed text zones in 94 noisy document images. The experimental results are shown in Table 4.5. All merging and splitting errors are counted as partially correct in the table. Before noise removal, ScanSoft has very poor results, an accuracy of 15.9%, on noisy documents under this metric. After analyzing the segmentation results, we found that ScanSoft tends to merge horizontally arranged zones into one zone, which is suitable for documents with simple layouts such as technical articles, but not for other document types such as business letters. The Docstrum algorithm outputs many more zones than ScanSoft, resulting in a higher accuracy (53.0%), but also a higher false alarm rate (114.1%). After noise removal, the accuracy of both algorithms increases significantly, from 15.9% to 48.4% for ScanSoft and from 53.0% to 78.0% for the Docstrum algorithm. The false alarm rate is reduced from 32.5% to 1.3% for ScanSoft and from 114.1% to 7.9% for Docstrum.

Fig. 4.9 shows the zone segmentation results for two noisy documents with the Docstrum algorithm before and after noise removal. The handwriting is output to another layer, not shown here. After noise removal, we see fewer splitting and

Table 4.5: Machine printed zone segmentation experimental results on 94 noisy document images (totally 1,374 zones), before and after noise removal.

	Before noise removal				After noise removal			
	Correctly segmented zones	False alarm zones	Partially correctly segmented zones	Missed zones	Correctly segmented zones	False alarm zones	Partially correctly segmented zones	Missed zones
ScanSoft	219 (15.9%)	446 (32.5%)	1,148 (83.7%)	7 (0.5%)	665 (48.4%)	18 (1.3%)	671 (48.8%)	38 (2.8%)
Docstrum	728 (53.0%)	1,568 (114.1%)	646 (47.0%)	0 (0.0%)	1,071 (78.0%)	109 (7.9%)	270 (19.7%)	33 (2.4%)

merging errors, and overall the segmentation results have significantly improved.

4.7 Summary and Future Work

In this chapter, we have presented an approach to segmenting and identifying handwriting from machine printed text in extremely noisy document images. Instead of using simple filtering rules, we treat noise as a distinct class. We use statistical classification techniques to classify each block into machine printed text, handwriting, and noise. We then use Markov random fields to incorporate contextual information for post-processing. Experiments show that MRFs are an effective tool for modeling local dependency among neighboring image components. After post-processing, the classification error rate is reduced by approximately 50%.

Our method is general enough to be extended to documents in some other scripts, such as Chinese, by re-training the classifier. However, our approach does not apply in a straightforward manner to cursive scripts such as Arabic. Two observations used to discriminate handwriting from machine printed text for English documents do not hold for Arabic documents. (1) Handwriting is more cursive than

97

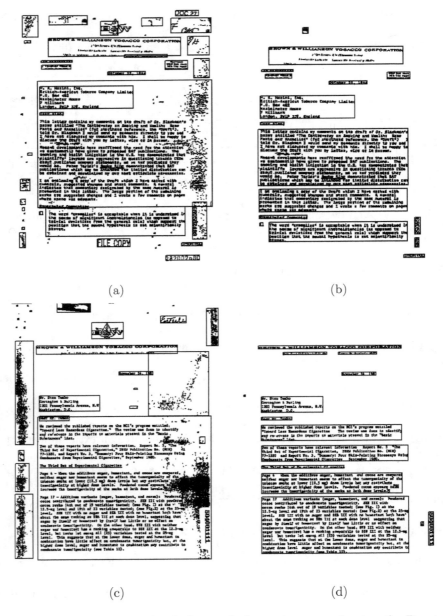

Figure 4.9: Zone segmentation before and after noise removal using the Docstrum algorithm. (a) and (c) show the results before noise removal. (b) and (d) are the results after noise removal.

machine printed text in English documents. However, machine printed Arabic text is cursive in nature. (2) People like to connect several characters during writing. However, many machine printed Arabic characters are also connected. Preliminary experiments using the same feature set proposed in this chapter were performed for Arabic documents, resulting in a low classification accuracy at word level. New features should be designed for Arabic documents. It should be much easier to distinguish handwriting from machine printed text in Arabic at the text line level. However, how to reliably extract text line from a heterogeneous and noisy document is challenging problem in itself.

Chapter 5

Handwriting Matching

5.1   Introduction

Handwriting samples (the same content) are often produced with large deformation by different persons or the same person at different times.[1] Due to the difficulty in re-producing a handwriting sample exactly, handwriting is often used to identify a person. On the other hand, large deformation is a challenge for other handwriting related applications such as handwriting recognition and retrieval. To study the deformation characteristics, it is crucial to automatically establish the correspondence between two handwriting samples, which have the following applications:

1. It is much easier to define similarity measures between handwriting samples after establishing the correspondence. The measures are often more robust and have more distinguishing power, comparing to the case where we do not know the correspondence [26, 105, 106].

2. The deformation characteristics learned can be used to synthesize visually realistic handwriting samples [107]. The quality of a trained statistical model for pattern classification depends highly on the quality of the training set.

---

[1] In this chapter, we exclude the variation produced by different contents (e.g., the difference between handwriting of letters 'a' and 'b'). Throughout this chapter, we focus on the deformation in shape.

As a rule, the more samples used for training, the higher the generalization capability of the trained classifier. Since collecting a large volume of handwriting samples is generally very expensive, synthesized samples are often used to enlarge the training set [3, 108].

In this chapter, we study the handwriting matching problem. In the next two chapters, we will apply our matching algorithm to handwriting synthesis and retrieval, respectively.

We study handwriting matching in a broader context of shape matching, which is often encountered in image analysis, computer vision, and pattern recognition. A shape may be represented as a set of features at different levels, such as points, line segments, curves, or surfaces. Shape matching may be performed on these representations. The survey paper by Loncaric [109] covers the extraction and representation of a shape. Different distance definitions between two features (i.e., point, lines, or curves) and their use in shape matching can be found in [110]. In general, the higher the level of a feature, the more difficult it is to extract the feature reliably. The extraction of interesting points, for example, is easy (sometimes trivial), and it is more general since lines and surfaces can be discretized as a set of points. Although such discretization is by no means optimal, reasonable matching results may be achieved in many cases [111]. Point matching, therefore, is often used in applications such as pose estimation [112, 113], medical image registration [114], surface registration [115, 116], object recognition [26], and handwriting recognition [105, 106].

In our approach, we first extract the handwriting skeleton, then uniformly sample a set of points from the skeleton. After that, we develop an algorithm to estimate the correspondence between two point sets. Compared to the original pixel-based representation (which can be seen as a representation with dense points), our approach demands fewer points. Several hundred points are enough to represent the structure of handwriting. Another advantage is our representation is more robust to stroke width variation, which is often introduced by the use of different writing tools or different digitization parameters (e.g., different parameters for scanning and binarization). Our point matching algorithm uses no or little prior knowledge of handwriting, and is general enough to be applied to other point pattern based nonrigid shape matching problems. In the remaining of this chapter, we describe our algorithm in the broader context of *nonrigid shape matching*, instead of the narrower *handwriting matching*.

5.1.1 Overview of Our Approach

Although the absolute distance between two points may change significantly under nonrigid deformation, the neighborhood structure of a point is generally well-preserved due to physical constraints. For example, a human face is a nonrigid shape, but the relative position of chin, nose, mouth, and eyes cannot deform independently due to underlying constraints of bones and muscles. These physical constraints restrict the deformation of the point set sampled from a face. The rough structure of a shape is typically preserved, otherwise even people cannot match shapes reliably under arbitrary large deformation. Such constraints may be represented as the ordering of points on a curve. Sebastian et al. [117] demonstrated the effectiveness of point ordering in matching curves, but for general shapes other than curves, local point ordering is difficult to describe, and is ignored in many point matching algorithms [26]. As a major contribution, we formulate point matching as an optimization problem to preserve local neighborhood structures. In addition to the physical constraint explanation, our approach is supported from cognitive experiments of human shape perception. Strong evidence suggests the early stages of human visual processing is local, parallel, and bottom-up, though feedback may be necessary in later stages. Preserving local neighborhood structures is important for people to detect and recognize shapes efficiently and reliably [118, 119].

In our approach, we formulate point matching as an optimization problem to preserve local neighborhood structures during matching. Our formulation has a simple graph matching interpretation, where each point is a node in the graph and two nodes are connected by an edge if they are neighbors. The optimal match between two graphs is the one that maximizes the number of matched edges (i.e., the number of neighborhood relations). Graph matching is an NP-hard problem. Exhaustive or branch-and-bound search for a global optimal solution is only realistic for graphs with few nodes. As an alternative, a discrete optimization problem can be converted to a continuous one, allowing several continuous optimization techniques to be applied [120, 121]. In our approach, we use the shape context distance to initialize graph matching, followed by a relaxation labeling process to refine the match.

5.1.2 Previous Work

Shapes can be roughly categorized as rigid or nonrigid, and the realization of a shape may undergo various deformations in captured images. With a small number of transformation parameters (six for a 2-D affine transformation), rigid shape matching under the affine [111, 115] or projective transformation [113] is relatively easy and has been widely studied with an extensive literature. Since it is impossible to sufficiently discuss previous publications without omitting many excellent works, we refer the reader to other survey papers for a large bibliography [122, 123]. Although many point matching algorithms developed for rigid shapes can tolerate some degree of noise or local distortions, large free-form deformation presents a significant challenge. Recently, point matching for nonrigid shapes has received more and more attention. In the following literature review, we will focus on publications on nonrigid shape matching.

Point matching for nonrigid shapes is problematic because the method must compensate for both linear distortions (i.e., translation, rotation, scale changes, and shear) and non-linear distortions. Therefore, the common framework of iterated correspondence and transformation estimation is used widely. The iterated closest point (ICP) algorithm is a well-known heuristic approach proposed by Besl and McKay [111, 115]. Assuming two shapes are roughly aligned, for each point in one shape, the closest point in the other shape is taken as the current estimate of the correspondence. The affine transformation estimated with the current correspondence will then bring two shapes closer. ICP was later extended for nonrigid free-form surfaces [116]. The framework consists of three stages. First, the rigid displacement is estimated using surface curvatures. Second, the global affine transformation is estimated using the ICP algorithm. Third, a local affine transformation (LAT) is attached to each point to deform the surface locally. Wakahara [105] used LAT to match and recognize handwritten characters. A dynamic window with a gradually decreasing size is used to estimate the local affine transformation for a point. This approach was later improved by combining global and local affine transformations to increase the robustness [106].

Although LAT has enough flexibility to model local nonrigid deformation, no standard exists to define the neighborhood window size to estimate the parameters of LAT. How to combine the global and local affine transformations is an open problem as well, so more flexible deformation models with closed-form representations are desired. In the literature on interpolation and approximation, radial basis functions (RBF) with different kernel functions, such as the thin plate spline (TPS) [124] and the Gaussian RBF [125], are widely used. Recently, the TPS deformation model began to be applied in point matching [26, 2] because it can be formulated as an optimal solution of the bending of a thin plate [124]. Chui and Rangarajan [2] proposed an optimization based approach, the TPS-RPM algorithm. The TPS model's bending energy and the average Euclidean distance between two point sets are combined in an objective function. The soft assignment technique and deterministic annealing algorithm are used to search for the optimal solution, which significantly outperforms the ICP algorithm on nonrigid point matching. Belongie et al. [26] proposed another method for nonrigid point matching. In this approach, a shape context is assigned to a given point, which describes the relative distribution of the other points in the shape. After defining the similarity between two points based on their shape contexts, the Hungarian algorithm [126] searches for the optimal match between the two point sets. Similarly, the TPS model brings two shapes closer in each iteration.

More recently, Glaunes et al. [127] proposed another point matching approach. Taking a point set as a sampling of the underlining distribution, they proposed a theory based on diffeomorphisms on distributions. Their formulation reduces to an optimization problem with a weighted summation of two parts: the energy associated with the deformation and the distance between two point sets under this deformation. This is similar to the objective function in [2], although no explicit deformation model is assumed. The variational method is used to search for the optimal deformation. Experimental results on synthesized data are encouraging, but more extensive tests should be performed to show the effectiveness of their approach.

Another interesting work is matching articulated objects [112]. An articulated object (such as a person) is composed of several rigid segments connected by pivot points. The deformation of rigid segments can be modeled with an affine transformation. A global hierarchical search strategy searches for and matches pivot points, and local matching of rigid segments is used to prune the search tree, thus reducing the computational cost [112].

The remainder of this chapter is organized as follows. In Section 5.2, we formulate point matching as an optimization problem. Our strategy to search for

106

an optimal solution is described in Section 5.3. Shape deformation models, such as the affine transformation and TPS, are discussed in Section 5.4, followed by a brief summary of our approach in Section 5.5. We demonstrate the robustness of our approach with experiments in Section 5.6, and the chapter concludes with a discussion of the future work in Section 5.7.

5.2   Problem Formulation

In this section, we formulate point matching as an optimization problem. Suppose a template shape T is composed of M points, $S_T = \{T_1, T_2, \cdots, T_M\}$, and a deformed shape D is composed of N points, $S_D = \{D_1, D_2, \cdots, D_N\}$. It is a common practice to enforce the one-to-one matching constraint in point matching, so the point sets S_T and S_D are augmented to $S'_T = \{T_1, T_2, \cdots, T_M, nil\}$ and $S'_D = \{D_1, D_2, \cdots, D_N, nil\}$ respectively, by introducing a dummy or *nil* point. A match between shapes T and D is $f : S'_T \Leftrightarrow S'_D$, where the matching of normal points is one-to-one, but multiple points may be matched to a dummy point.

Under a rigid transformation (i.e., translation and rotation), the distance between any pair of points is preserved. Therefore, the optimal match $\hat{f}$ is

$$\hat{f} = arg \min_f C(T, D, f), \tag{5.1}$$

where

$$
\begin{aligned}
C(T, D, f) = \;& \sum_{m=1}^{M} \sum_{i=1}^{M} \left(\|T_m - T_i\| - \|D_{f(m)} - D_{f(i)}\| \right)^2 \\
& + \sum_{n=1}^{N} \sum_{j=1}^{N} \left(\|D_n - D_j\| - \|T_{f^{-1}(n)} - T_{f^{-1}(j)}\| \right)^2 .
\end{aligned}
\tag{5.2}
$$

In this cost function, we penalize any matching error which does not preserve the distance of a point pair. If $M = N$ and no points are matched to dummy points, the first term and the second term in (5.2) should be equal, and the optimal match should achieve zero penalty, $C(T, D, \hat{f}) = 0$. Points matching a dummy point need special treatment, and to simplify the representation, we do not describe such treatment here. We will return to this issue.

If nonrigid deformation is present, the distance between a pair of points will not be preserved, especially for points which are far apart. On the other hand, due to physical constraints and in order to preserve the rough structure, the local neighborhood of a point may not change freely. We, therefore, define a neighborhood for point i as $\mathcal{N}_i$. The neighborhood relationship is symmetric, meaning if $j \in \mathcal{N}_i$ then $i \in \mathcal{N}_j$. Since we want to preserve the distances of neighboring point pairs under deformation, (5.2) becomes

$$
\begin{aligned}
C(T, D, f) \;=\; & \sum_{m=1}^{M} \sum_{i \in \mathcal{N}_m} \left(\|T_m - T_i\| - \|D_{f(m)} - D_{f(i)}\| \right)^2 \\
& + \sum_{n=1}^{N} \sum_{j \in \mathcal{N}_n} \left(\|D_n - D_j\| - \|T_{f^{-1}(n)} - T_{f^{-1}(j)}\| \right)^2 . \quad (5.3)
\end{aligned}
$$

The absolute distance of a pair of points is not preserved well under scale changes. Therefore, we quantize the distance to two levels as

$$
\|T_m - T_i\| = \begin{cases} 0 & m \in \mathcal{N}_i \\ 1 & m \notin \mathcal{N}_i \end{cases} \quad \text{and} \quad \|D_n - D_j\| = \begin{cases} 0 & n \in \mathcal{N}_j \\ 1 & n \notin \mathcal{N}_j \end{cases} . \quad (5.4)
$$

(5.3) then is simplified to

$$
C(T, D, f) = \sum_{m=1}^{M} \sum_{i \in \mathcal{N}_m} d(f(m), f(i)) + \sum_{n=1}^{N} \sum_{j \in \mathcal{N}_n} d(f^{-1}(n), f^{-1}(j)), \quad (5.5)
$$

where

$$d(i,j) = \begin{cases} 0 & j \in \mathcal{N}_i \\ 1 & j \notin \mathcal{N}_i \end{cases}. \tag{5.6}$$

To deal with points matched to a dummy point, we let $d(.,nil) = d(nil,.) = d(nil,nil) = 1$ to discourage the match.

In the following, we rewrite the objective function of (5.5), and interpret it as a graph matching problem. First, we subtract a constant term from $C(T,D,f)$.

$$
\begin{aligned}
C'(T,D,f) &= C(T,D,f) - \sum_{m=1}^{M}\sum_{i\in\mathcal{N}_m} 1 - \sum_{n=1}^{N}\sum_{j\in\mathcal{N}_n} 1 \\
&= \sum_{m=1}^{M}\sum_{i\in\mathcal{N}_m} [d(f(m),f(i)) - 1] + \sum_{n=1}^{N}\sum_{j\in\mathcal{N}_n} [d(f^{-1}(n),f^{-1}(j)) - 1] \\
&= -\sum_{m=1}^{M}\sum_{i\in\mathcal{N}_m} \delta(f(m),f(i)) - \sum_{n=1}^{N}\sum_{j\in\mathcal{N}_n} \delta(f^{-1}(n),f^{-1}(j)) \tag{5.7}
\end{aligned}
$$

where

$$\delta(i,j) = 1 - d(i,j) \tag{5.8}$$

Minimizing $C(T,D,f)$ is equivalent to minimizing $C'(T,D,f)$ since the difference between them is a constant. Therefore, the minimization problem of (5.1) is equivalent to the following maximization problem.

$$\hat{f} = arg\max_{f} S(T,D,f) \tag{5.9}$$

where

$$S(T,D,f) = \sum_{m=1}^{M}\sum_{i\in\mathcal{N}_m} \delta(f(m),f(i)) + \sum_{n=1}^{N}\sum_{j\in\mathcal{N}_n} \delta(f^{-1}(n),f^{-1}(j)) \tag{5.10}$$

This formulation has a simple graph matching interpretation. Each point is a node in the graph, and two nodes are connected by an edge if they are neighbors. The

109

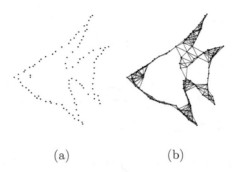

(a)   (b)

Figure 5.1: A point set (a) and its graph representation (b).

dummy node is not connected to other nodes in the graph. If connected nodes m and i in one graph are matched to connected nodes $f(m)$ and $f(i)$ in the other graph, $\delta(f(m), f(i)) = 1$. The optimal solution of (5.9) is the one that maximizes the number of matched edges of two graphs.

No obvious neighborhood definition exists for a point set. In the following we present a simple neighborhood definition.[2] Initially, the graph is fully connected, then we remove long edges until a pre-defined number of edges are preserved. Supposing M nodes in the graph and on average each point has E_{ave} neighbors, then the number of preserved edges is $M \times E_{ave}/2$ ($E_{ave} = 5$ in default). With this neighborhood definition, the graph representation of a point set is translation, rotation, and scale invariant. Fig. 5.1 shows a point set and its graph representation. We expect points connected with an edge move together under deformation, so the structure of the graph is preserved.

---

[2]Our framework is general enough to incorporate other neighborhood definitions. Please refer to Section 5.6 for a definition, which is robust under nonuniform scale changes for different parts of a shape.

Graph matching (more generally, attributed relational graph matching) is used in [128] and [129] to match road maps extracted from aerial photographs. Our graph definition differs from theirs, where road intersections are nodes in the graph and two nodes are connected by an edge if a road appears between two intersections. Such a graph definition is natural for a road map, but errors in road detection will change the graph structure. In our case, given an arbitrary set of points, there is no such natural definition of connections among points. Graph matching is widely used in many fields such as computer vision and pattern recognition. There are various kinds of graph structures, and many different metrics are available in the literature to evaluate a match between two graphs [130]. Our graph representation and the corresponding matching metric are derived from the observation (or assumption) that nonrigid deformation will not change the neighborhood of a point significantly.

5.2.1 Matching Matrix

We can represent the matching function f in (5.10) with a set of supplemental variables, which can be organized as a matrix P with dimension $(M+1) \times (N+1)$.

$$
P =
\left[
\begin{array}{ccc|c}
p_{11} & \cdots & p_{1N} & p_{1,nil} \\
\vdots & \vdots & \vdots & \vdots \\
p_{M1} & \cdots & p_{MN} & p_{M,nil} \\
\hline
p_{nil,1} & \cdots & p_{nil,N} & 0
\end{array}
\right]
\tag{5.11}
$$

If point T_m in the template shape T is matched to point D_n in the deformed shape D, then $P_{mn} = 1$; otherwise $P_{mn} = 0$. The last row and column of P represent the case that a point may be matched to a dummy point. Matrix P satisfies the following normalization conditions

$$\sum_{n=1}^{N+1} P_{mn} = 1 \quad \text{for } m = 1, 2, \cdots, M, \tag{5.12}$$

$$\sum_{m=1}^{M+1} P_{mn} = 1 \quad \text{for } n = 1, 2, \cdots, N. \tag{5.13}$$

Using matrix P, the objective function (5.10) can be written as

$$S(T, D, P) = 2 \sum_{m=1}^{M} \sum_{i \in \mathcal{N}_m} \sum_{n=1}^{N} \sum_{j \in \mathcal{N}_n} P_{mn} P_{ij} \tag{5.14}$$

5.3 Searching for an Optimal Solution

Since $P_{mn} \in \{0, 1\}$, searching for an optimal P that maximizes $S(T, D, P)$ is a difficult discrete combinatorial problem. In our approach, we use relaxation labeling to solve the optimization problem, where the condition $P_{mn} \in \{0, 1\}$ is relaxed as $P_{mn} \in [0, 1]$ [121]. After relaxation, P_{mn} is a real number, and the problem is converted to a constrained optimization problem with continuous variables.

5.3.1 Matching Initialization

The performance of relaxation labeling depends heavily on the initial value of the matching probability matrix P. We need a good initial measure of the matching probabilities. One option involves assigning an attribute, such as the color or intensity gradients of the pixel, to a point if it is extracted as a pixel in an image [131]. We can then compute the similarity between a pair of points, and convert it to a measure of the matching probability. If a set of points is given without any additional information, the shape context provides an effective way to compute the

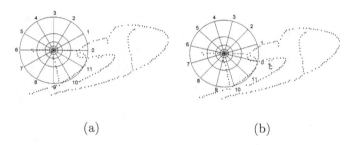

(a)   (b)

Figure 5.2: Shape context of a point. (a) Basic shape context. (b) Our rotation invariant shape context. The point labeled with * is the mass center of the point set.

similarity between two points [26]. In our approach, we use the shape context distance to initialize the point matching probabilities. If other attributes of a point are available, they can be easily incorporated into our framework.

To extract the shape context of a point, an array of bins is placed around the point, as shown in Fig. 5.2a. The number of points inside each bin is calculated as the context of this point. Therefore, the shape context of a point is a measure of the distribution of other points relative to it. Bins that are uniform in log-polar space are used to make the descriptor more sensitive to positions of nearby points than to those of points far away. Five bins for the radius and 12 bins for the rotation angle are used throughout our experiments. Consider two points, m in one shape, and n in the other shape. Their shape contexts are $h_m(k)$ and $h_n(k)$, for $k = 1, 2, \ldots, K$, respectively. Let C_{mn} denote the cost of matching these two points. As shape contexts are distributions represented as histograms, it is natural to use the χ^2 test statistic [26]

$$C_{mn} = \frac{1}{2} \sum_{k=1}^{K} \frac{[h_m(k) - h_n(k)]^2}{h_m(k) + h_n(k)} \qquad (5.15)$$

113

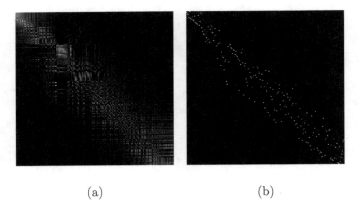

(a)   (b)

Figure 5.3: Point matching probability matrix P. The matching probabilities to a dummy point are not shown. White represents a high probability. (a) Initial probabilities using the shape context distance. (b) After 300 iterations of relaxation labeling updates.

The Gibbs distribution is widely used in statistical physics and image analysis to relate the energy of a state to its probability [69]. Taking the cost C_{mn} as the energy of the state that points m and n are matched, the probability of the match is

$$P_{mn} \propto e^{-C_{mn}/T_{init}} \tag{5.16}$$

Parameter T_{init} is used to adjust the reliability of the initial probability measures, where $T_{init} \in [0.05, 0.1]$ is appropriate according to our experiments. We set the probability for a point matching to a dummy point, $P_{m,nil}$ or $P_{nil,n}$, to 0.2. Experiments show that our approach is not sensitive to this parameter. Fig. 5.3a shows the initial matching probability matrix P of two shapes.

Obviously, shape context is translation invariant. Using bin arrays with an adaptive size according to the mean point distance of a shape, the shape context is

114

scale invariant too [26], but it is sensitive to large rotations. In some applications, rotation invariance is required. Our graph representation is rotation invariant, so we need a rotation invariant initialization scheme. A complete rotation invariant shape context was proposed using the tangent direction at each point as the positive x-axis for the local coordinate system [26]. One drawback of this approach is that the tangent direction, defined for gray-scale images, does not apply for binary images. Furthermore, if only the point set is given without accessing the original image, we cannot estimate the tangent direction. Another drawback is that as a first-order derivative operation, the estimate of the tangent direction is sensitive to noise. Instead, in our approach, we use the mass center of a point set as a reference point and use the direction from a point to the mass center as the positive x-axis for the local coordinate system. Fig. 5.2b shows our rotation invariant shape context. If there is zero mean white noise in point position measurements, after averaging, the effect of noise to the mass center is reduced. Therefore, our approach is more robust than the tangent direction based approach under noise.

5.3.2 Relaxation Labeling

Relaxation labeling was first proposed in a seminal paper by Rosenfeld et al. in the mid-1970s [121]. The basic idea is to use iterated local context updates to achieve a globally consistent result. The contextual constraints are expressed in the compatibility function $R_{mn}(i, j)$, which, in our case, measures the strength of compatibility between T_m matching D_n and T_i matching D_j. The support function S_{mn} measures the overall support the match between points T_m and D_n receives from its neighbors.

$$S_{mn} = \sum_{i \in \mathcal{N}_m} \sum_{j \in \mathcal{N}_n} R_{mn}(i,j) P_{ij} \qquad (5.17)$$

The original updating rule is[3]

$$P_{mn} := \frac{P_{mn} S_{mn}}{\sum_{j=1}^{N} P_{mj} S_{mj}}. \qquad (5.18)$$

The denominator enforces the normalization constraint. Traditionally, only the one-way normalization constraint, Eq. (5.12), is enforced in relaxation labeling.

In the original paper, S_{mn} is defined heuristically. Although with ad hoc heuristic arguments, a variety papers later reported on the practical usefulness of the algorithm (see [132] for a review and an extensive bibliography). The success in real applications and the heuristic flavor of the algorithm motivated investigators to establish a theoretic foundation. There are two approaches. Some have tried to set the labeling problem within a probabilistic framework using Bayesian analysis [133, 128]. The Bayesian theory can explain only one iteration of the relaxation process. An alternative explicitly defines some quantitative measure of consistency to be maximized, then formulates the labeling problem as one of optimization [134, 135]. Projected gradient methods are often used to optimize the objective function. In these theories, the support function S_{mn} is defined as the derivative of the objective function with respect to P_{mn} [135]. The updating rule of the projected gradient methods is

$$P := P + \gamma Q(S) \qquad (5.19)$$

---

[3]In the original paper, the support function S is defined in a heuristic way in the range of $[-1, 1]$. In order to satisfy $P \geq 0$ after updating, $1 + S$ is used to substitute S in both the numerator and denominator in the updating rule [121].

where γ is the updating step and S is a matrix composed of elements S_{mn}. $Q(S)$ is a projection operation of S to limit the range of P_{mn} to $[0, 1]$ and enforce normalization constraints. In the case of boundary points (i.e., having at least one component of the probabilities equal to zero or one), the projection operation is much more complicated and the procedure becomes computationally expensive. Furthermore, the updating step γ is difficult to tune. An increase in the objective function is guaranteed only when infinitesimal steps are taken, and searching for the optimal step size in each iteration is computationally expensive. Recently, Pelillo [136] showed that the original updating rule in (5.18) does converge to a local minimum if 1) the objective function is a polynomial with nonnegative coefficients, and 2) S_{mn} is defined as a gradient of the objective function. The advantages of this updating rule, compared with the projected gradient methods, are 1) computationally expensive projection operations are avoided, and 2) it is parameter free. We tried several updating rules compared in [137] and found that the updating formula of (5.18) is robust and achieves better results. With our objective function of (5.14), S_{mn} takes the form of

$$S_{mn} = 4 \sum_{i \in \mathcal{N}_m} \sum_{j \in \mathcal{N}_n} P_{ij} \qquad (5.20)$$

Since $S_{mn} \geq 0$, the constraint that $P_{mn} \in [0, 1]$ is satisfied after normalization. Interpreted in the relaxation labeling theory, our compatibility function is $R_{mn}(i, j) = 1$ if a pair of neighbors T_m and T_i are matched to a pair of neighbors D_n and D_j; otherwise $R_{mn}(i, j) = 0$. We can easily show that our objective function corresponds to the average local consistency measure of the Hummel-Zucker consistency

117

theory [135]. According to the consistency theory, each update step will increase the overall consistency of the system. Each unambiguous consistent solution is a local attractor. Starting from a nearby point, the relaxation labeling process will converge to it [135, 136]. Although there is no guarantee that the updating process will converge to an unambiguous solution starting from an *arbitrary* initialization, our experiments show that most elements of matrix P do converge to zero or one.

In previous applications of the relaxation labeling technique, a many-to-one match is allowed [138, 139, 140, 141, 142]. Only a one-way normalization constraint, either (5.12) or (5.13), is enforced. Unfortunately, in many applications, one-to-one match is desired. Projected gradient methods may be modified to enforce one-to-one match. The projection operation, however, is computationally expensive, and it is unclear how to find a projection satisfying two-way normalization constraints. In our approach, a different approach based on alternated row and column normalizations of the matching probability matrix P is used to enforce one-to-one match [2]. A nonnegative square matrix with each row and column summing to one is called a doubly stochastic matrix. Sinkhorn [143] showed that the iterative process of alternated row and column normalizations will convert a matrix with positive elements to a doubly stochastic matrix. The conclusion can be extended to a non-square matrix with positive elements. We call a matrix where each row and column (except the last row and column) sums one a generalized doubly stochastic matrix. We can show that alternated row and column normalizations (except the last row and column) of a matrix with positive elements will result in a generalized doubly stochastic matrix. Please refer to the Appendix of this chapter for the proof. This technique was also used in the soft assignment point matching approach without proof [2]. Though relaxation labeling with one-way normalization constraint can be theoretically well founded [136], it is not clear whether the updating process will converge to a local optimum and increase the consistency after imposing the one-to-one matching constraint. We found the updating process still converges through experiments, but we cannot prove it theoretically.

Fig. 5.3a shows the initial value of the point matching probability matrix P of two shapes. After each relaxation labeling update, we perform alternated row and column normalizations to matrix P. Generally, a few rounds will bring a matrix close to a generalized doubly stochastic matrix. After 300 iterations of relaxation labeling updates, the ambiguity of matches decreases. As shown in Fig. 5.3b, most elements of the matrix converge to zero or one.

After relaxation labeling updates, points with maximum matching probability less than P_{min} ($P_{min} = 0.95$) are labeled as outliers by matching them to dummy points. The matched point pairs are used to estimate the parameters of the affine or TPS deformation model, and the estimated parameters are used to transform the template shape to bring it closer to the deformed shape. In some application scenarios (e.g., the experiments in Section 5.6.1), we may want to find as many matches as possible. Unfortunately, the ratio of points matched to dummy points by the relaxation labeling updates cannot be controlled directly. After several iterations of correspondence and transformation estimations, two point sets may be close to each other. Therefore, in the last round, we find the optimal one-to-one match by minimizing the summation of Euclidean distances from the transformed template shape to the deformed shape.

$$Dist = \sum_{m=1}^{M} \|T_m^* - D_{f(m)}\| \tag{5.21}$$

where T_m^* is a point from the template shape after TPS transformation. The optimal
match $\hat{f}$ of (5.21) can be found using the Hungarian algorithm [126]. We emphasize
that the above process is necessary only to disable the automatic outlier rejection
scheme of the algorithm and find as many matches as possible between two point
sets.

5.3.3 Relationship to Previous Work

One difference from the previous applications of relaxation labeling on point matching [142] is that we use it to solve a constrained optimization problem so the relaxation labeling process is guaranteed to converge to a local optimal solution [136].
In the previous work, relaxation labeling is used in an ad hoc way without an objective function to be optimized, so no clear indication exists for the quality of the
solution. Furthermore, unlike previous work, we can enforce one-to-one matching
in our approach if necessary.

The relaxation labeling method used in our approach is similar to the well-
known soft assignment technique [120, 2]. Both convert the discrete combinatorial
optimization problem to one with continuous variables by assigning a probability
measure to a match. The procedure is called *relaxation* or *soft* in these two techniques respectively. Generally, deterministic annealing is used to solve the soft
assignment problem. It begins with a high temperature where the matching proba-

bilities are uniform. By gradually decreasing the temperature, the matching proba-bilities will converge to a local optimal solution of the objective function. An appro-priate choice of the initial temperature and temperature reduction ratio is necessary to achieve good results [2]. On the contrary, the relaxation labeling based approach is parameter free. Another advantage is that we can easily incorporate a mean-ingful initialization in relaxation labeling. Since the distribution of local optima is complex, a good initialization is crucial to achieve a good result. Unfortunately, it is difficult to incorporate an initialization method into the deterministic anneal-ing framework. It is also a drawback of another continuous optimization technique for graph matching proposed by Pelillo [144]. We tested the soft assignment based graph matching method [120] and found the results were worse than the relaxation labeling based approach.

5.4 Shape Deformation Models

It is difficult to achieve a good match for shapes under both rigid and nonrigid dis-tortions with a single-step approach. The strategy of iterated point correspondence and transformation estimations is widely used for nonrigid shape matching. In our approach, for the first iteration, the affine transformation between two shapes is es-timated and corrected. Instead of using the least squares (LS) estimator to estimate parameters of the affine transformation [26], we use a more robust least median squares (LMS) estimator. In the following iterations, the thin plate spline (TPS) deformation model is exploited to bring two shapes closer. Our approach is similar to [26] except that a more robust LMS estimator is used to estimate the affine

transformation, instead of the LS estimator.

5.4.1 Affine Transformation Estimation Based on LMS

The LS estimator is widely used to estimate transformation parameters. Suppose

point (x_i, y_i) is matched to point (u_i, v_i), for $i = 1, 2, \cdots, n$, the optimal parameters

of the affine transformation minimize the summation of squares of the regression

errors.

$$\hat{A}, \hat{T} = arg \min_{A,T} \sum_{i=1}^{n} \left\| \begin{pmatrix} u_i \\ v_i \end{pmatrix} - A \begin{pmatrix} x_i \\ y_i \end{pmatrix} - T \right\|^2 \tag{5.22}$$

where A is a 2×2 matrix representing the rotation and anisotropic scale changes,

and T is a translation vector. One advantage of the LS estimator is that closed-form

solutions are available [145]. It is, however, sensitive to outliers in matching [146].

The breakdown point is often used to evaluate robustness of an estimator under

outliers, which is defined as the smallest proportion of observations that must be

replaced by arbitrary values in order to force the estimator to produce values arbi-

trarily far from the true values [147]. The breakdown point of the LS estimator is

0%. Furthermore, it is difficult to detect outliers based on the regression residual

errors since they may spread over all of the points [146].

In general, the results of the first iteration of point matching may be noisy

with many errors, so a more robust estimator is required. Several robust regression

methods have been proposed in the statistics literature. Among them, the least

median squares (LMS) estimator achieves the highest possible break down point,

50% [146]. Instead of minimizing the summation of squares of regression errors, the LMS estimator minimizes the median of the regression errors.

$$
\hat{A}, \hat{T} = arg \min_{A,T} \ \text{median} \left\{ \left\| \begin{pmatrix} u_i \\ v_i \end{pmatrix} - A \begin{pmatrix} x_i \\ y_i \end{pmatrix} - T \right\|^2 \quad \text{for } i = 1, 2, \cdots, n \right\}
$$
(5.23)

There are no closed-form solutions for (5.23). Normally, we randomly select a subset with three matched pairs (which can determine an affine transformation) and calculate the median of the regression errors using the estimated parameters. Iterating the random selection procedure, an optimal solution of (5.23) can be achieved. Suppose there are n matched pairs and about 50% of them are wrong. In the worse case, we must select at least $\begin{pmatrix} n \\ 3 \end{pmatrix} - \begin{pmatrix} n/2 \\ 3 \end{pmatrix} + 1$ different subsets to ensure at least one subset without outliers is selected. This is too pessimistic. In real applications, we only need to examine a small number of subsets. After examining k subsets, the probability of having at least one good subset is $1 - \left[1 - \begin{pmatrix} n/2 \\ 3 \end{pmatrix} / \begin{pmatrix} n \\ 3 \end{pmatrix} \right]^k$ (assuming sampling with replacement). For example, let $n = 200$, the probability of getting at least one good subset in 50 random selections is 99.8%. The LMS estimator can be used to estimate the affine transformation without knowing the correspondence between two point sets [148]. Without rough correspondence, a large number of subsets need to be examined.

5.4.2 TPS Deformation Model

The TPS model is often used for representing flexible coordinate transformations because it has a physical explanation and closed-form solutions in both transformation and parameter estimation [124]. It has been used in nonrigid shape matching in [26] and [2]. Two TPS models are used for the 2-D coordinate transformation. Suppose point (x_i, y_i) is matched to (u_i, v_i) for $i = 1, 2, \cdots, n$, let $z_i = f(x_i, y_i)$ be the target function value at location (x_i, y_i), we set z_i equal to u_i and v_i in turn to obtain one continuous transformation for each coordinate. The TPS interpolant $f(x, y)$ minimizes the bending energy

$$I_f = \int\int_{\mathcal{R}^2} \left(\frac{\partial^2 f}{\partial x^2}\right)^2 + 2\left(\frac{\partial^2 f}{\partial x \partial y}\right)^2 + \left(\frac{\partial^2 f}{\partial y^2}\right)^2 dxdy \tag{5.24}$$

and has the solution of the form

$$f(x, y) = a_1 + a_x x + a_y y + \sum_{i=1}^{n} w_i U(\|(x_i, y_i) - (x, y)\|) \tag{5.25}$$

where $U(r)$ is the kernel function, taking the form of $U(r) = r^2 log r^2$. The parameters of the TPS models w and a are the solution of the following linear equation

$$\begin{bmatrix} K & P \\ P^T & 0 \end{bmatrix} \begin{bmatrix} w \\ a \end{bmatrix} = \begin{bmatrix} z \\ 0 \end{bmatrix} \tag{5.26}$$

where $K_{ij} = U(\|(x_i, y_i) - (x_j, y_j)\|)$, the ith row of P is $(1, x_i, y_i)$, w and z are column vectors formed from w_i and z_i respectively, and a is the column vector with elements a_1, a_x, and a_y.

If errors appear in the matching results, we use regularization to trade off between exact interpolation and minimizing the bending energy as follows.

$$H_f = \sum_{i=1}^{n} [z_i - f(x_i, y_i)]^2 + \lambda I_f \qquad (5.27)$$

where λ is the regularization parameter, controlling the amount of smoothing. The regularized TPS can be solved by replacing K in (5.26) with $K + \lambda I$, where I is the $n \times n$ identity matrix [149, 150]. We set $\lambda = 1$ in the following experiments.

5.5 Summary of Our Approach

Following is a brief summary of our approach.

Input: Two point sets, $T_1, T_2, \ldots, T_M$ from the template shape T, and $D_1, D_2, \ldots, D_N$ from the deformed shape D.

Output: The correspondence between two point sets.

1. Set the transformed template shape T^* as T.

2. Set iteration number to one.

3. Calculate the shape context for each point in T^* and D, and use (5.15) to calculate the distance between each point pair T_m^* and D_n.

4. Use (5.16) to initialize the matching probability matrix P and convert it to a generalized doubly stochastic matrix by alternated row and column normalizations.

5. Use (5.18) to update the matching probability matrix R ($R = 300$) times. After each update, convert matrix P to a generalized doubly stochastic matrix.

125

6. If the iteration number is one, use LMS to estimate the affine transformation between T and D.

7. Otherwise, use (5.26) to estimate parameters of the TPS deformation model between T and D.

8. Transform template point set T to T^* using the estimated deformation parameters.

9. Increase the iteration number by one. If the iteration number is less than I_{max} ($I_{max} = 10$), go to step 3.

Suppose both shapes have N points, the computation cost of shape context distances is in the order of $O(N^2)$. Relaxation labeling updates will take $O(N^2)$ time. The computational complexity of the algorithm may be largely dependent on the implementation of the spline deformation, which can be $O(N^3)$ in the worst case. With our un-optimized C++ implementation, matching two shapes (each with 100 points) takes about 1.6 seconds on a PC with a 2.8 GHZ CPU.

5.6 Experiments

In this section, we show our approach preserves sequential ordering of points (a degenerated neighborhood structure) on open curves and closed contours during matching. We also test our approach in matching real handwriting samples. We then quantitatively compare it with two state-of-the-art algorithms for robustness under deformation, noise in point locations, outliers, occlusion, and rotation.

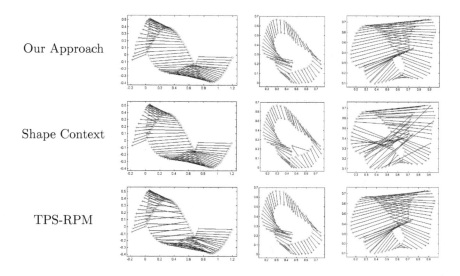

Figure 5.4: Point matching results on open curves (the left column) and closed contours (the middle and right columns). Top row: our approach. Middle row: the shape context algorithm. Bottom row: the TPS-RPM algorithm.

5.6.1 Some Examples

We have tested our algorithm on the samples used in [2] and compared our results with two other algorithms: shape context [26] and TPS-RPM [2]. In these examples, the template and deformed shapes have the same number of points. To achieve a direct and fair comparison, we prefer to match as many point pairs as possible without rejection. The shape context algorithm can achieve this by setting the outlier ratio to zero. The TPS-RPM algorithm and our relaxation labeling based approach may reject some points as outliers by matching them to a dummy point. Unfortunately, there are no parameters available in either algorithm to adjust the ratio of rejected points explicitly. In this experiment, after point matching and shape transformation are finished, we use the approach discussed in Section 5.3.2 to minimize the summation of Euclidean distances between the transformed template point set and the deformed point set (see, Equation (5.21)).

Fig. 5.4 shows the point matching results of three algorithms on a pair of open curves and two pairs of closed contours. As shown in the left column, all three algorithms achieve good results for the pair of open curves even though the deformation is large. Neighboring points may swap their matches in TPS-RPM. For the first pair of closed contours, all algorithms achieve reasonable results, but the shape context algorithm introduces a few mismatches as shown in the middle column of Fig. 5.4. Since the rotation between two shapes is large for the second pair of closed contours, the rotation invariance shape context is used for initialization in our approach and the shape context algorithm. Both our approach and TPS-RPM achieve good results and preserve the sequential ordering of points. The result of the shape context algorithm is not as good: neighboring points in one shape may be matched to points far apart in the other shape.

We also test our algorithm for handwriting matching. Figs. 5.5a and b show two samples of handwritten initials from the same person. We notice the structural change for handwriting is large: the characters overlap each other in the first sample, but they are well separated in the second sample. We uniformly sample 200 points from the skeletons of the handwriting, as shown in Figs. 5.5c and d. Fig. 5.5e shows the point matching results using our approach. Points labeled with green color are outliers rejected by our algorithm. On the D's, most points are assigned with correct correspondence. The touching parts of the S are assigned with low matching

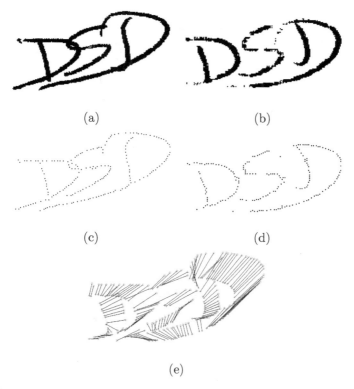

(a)   (b)

(c)   (d)

(e)

Figure 5.5: Handwriting matching. (a) and (b) two handwritten initials from the same person. (c) and (d) the point sets (each with 200 points) sampled from the skeletons of (a) and (b), respectively. (e) Point matching results using our approach.

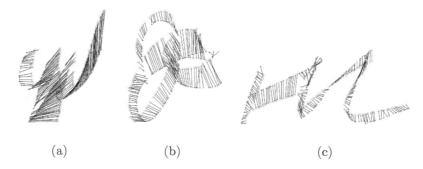

(a)   (b)   (c)

Figure 5.6: More examples of handwriting matching using our approach.

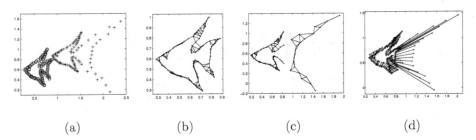

(a)   (b)   (c)   (d)

Figure 5.7: A neighborhood definition which is robust under large nonuniform scale changes for different parts of a shape. (a) Point sets of the template (o) and deformed (+) shapes. (b) Template graph with 210 edges. (c) Deformed graph with 196 edges. Among them, 178 (91%) edges also present in the template graph. (d) Point matching result.

probabilities, therefore rejected as outliers. More examples of handwriting matching are shown in Fig. 5.6.

Our approach is general enough to incorporate other neighborhood definitions. In this experiment, we use a neighborhood definition which is robust when different parts of a shape have significantly different scales. For two points in a shape, they are neighbors if and only if they are both in each other's top E_{ave} nearest points. This neighborhood relationship is still symmetric and scale invariant. A fish shape is used to synthesize test samples. First, we applied a moderate amount of nonrigid deformation to the template shape, and then enlarged the fish tail. Fig. 5.7b and c show the graphs of the template and deformed shapes when $E_{ave} = 5$ and the fish tail is enlarged to four times of the original size. This neighborhood definition is robust: 91% of the edges in the deformed graph have a correspondence in the template graph, and a good point match is achieved as shown in Fig. 5.7d.

130

5.6.2 Quantitative Evaluation

Synthetic data is easy to obtain and can be designed to test specific aspects of an algorithm. We test our algorithm on the same synthesized data as in [2] and [26]. Three sets of data are designed to measure the robustness of an algorithm under **deformation, noise** in point locations, and **outliers**. In each test, the template point set is subjected to one of the above distortions to create a *target* point set (for the latter two test sets, a moderate amount of deformation is present). Two shapes (a fish and a Chinese character) are used, and 100 samples are generated for each degradation level. We then run our algorithm to find the correspondence between these two sets of points and use the estimated correspondence to warp the template shape. The accuracy of the match is quantified as the average Euclidean distance between a point in the warped template and the corresponding point in the target. Alternative evaluation metrics are possible (e.g., the number of correctly matched point pairs), but in order to compare our results directly with two other algorithms, we use the same evaluation metric as in [2] and [26]. Fig. 5.8 shows several examples from the synthesized data sets, and Fig. 5.9 demonstrates the quantitative evaluation results. Since the new neighborhood definition presented above is not robust to outliers, the original version is used throughout this experiment. Our algorithm performs best on the deformation and noise sets. For the outlier test set, however, no clear winner appears. The TPS-RPM algorithm outperforms our algorithm on the Chinese character shape under large outlier ratios. Since points are spread out on the Chinese character shape, when a large number of outliers are present, the

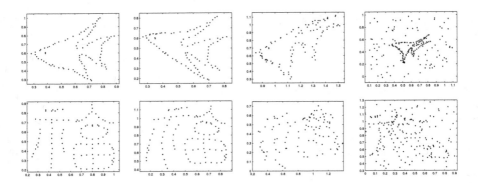

Figure 5.8: Chui-Rangarajan synthesized data sets. The template point sets are shown in the first column. Column 2-4 show examples of target point sets for the deformation, noise, and outlier tests respectively.

neighborhood of a point changes significantly (as shown in the last column of Fig. 5.8), which violates our assumption. Points on the fish shape are clustered, and the neighborhood of a point is preserved well even with a large outlier ratio. Therefore, better results are achieved by our algorithm on this shape.

Often present in real applications, occlusion is a challenge for many algorithms. In the following experiments, we test the three algorithms under occlusion using synthesized data. A moderate amount of nonlinear deformation is applied to a shape. We then randomly select a point and remove it with some of its closest points. Six occlusion levels are used: 0%, 10%, 20%, 30%, 40%, and 50%, and 100 samples are generated for each level. The top row of Fig. 5.10 shows two synthesized samples. Quantitative evaluation results are shown in the bottom row of Fig. 5.10. The TPS-RPM algorithm treats all extra points as outliers, which are assumed to be independently distributed. Since it does not model the distribution of occlusions

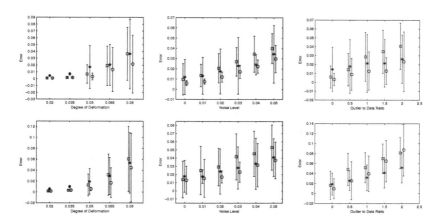

Figure 5.9: Comparison of our results (○) with the TPS-RPM (*) and shape context (□) algorithms on the Chui-Rangarajan synthesized data sets. The error bars indicate the standard deviation of the error over 100 random trials. Top row: the fish shape. Bottom row: the Chinese character shape.

well, the performance of TPS-RPM deteriorates quickly. In our approach, the change of neighborhood structure is restricted to points close to the occlusion. As shown in Fig. 5.10, our approach achieves the best results for up to 40% occlusion. When the occlusion ratio is large, a shape is likely to be broken into several parts and the neighborhood structure of almost all remaining points may be changed. Therefore, when the occlusion ratio is 50%, the difference between our approach and the shape context algorithm is small.

In some applications, rotation invariance is a critical property of a shape matching algorithm. We test our algorithm under rotations using synthesized data of the same fish and Chinese character shapes. A moderate amount of nonlinear deformation is applied to a shape, and the ground-truthed correspondences are used to correct the rotation introduced in the deformation. We then rotate the deformed shape. The probability of selecting a clockwise or counterclockwise rotation is equal. Six rotations are used: 0, 30, 60, 90, 120, and 180 degrees. One hundred samples are generated for each rotation. The top row of Fig. 5.11 shows two synthesized samples. At the first iteration, the rotation invariant shape context distance is used to initialize the matching probabilities in our approach. The rotation between two shapes is corrected by the affine transformation in the first iteration. After that, the normal shape context distance is used. Quantitative evaluation results appear in the bottom row of Fig. 5.11. We can see that our method is truly rotation invariant, and it consistently outperforms the shape context algorithm. TPS-RPM, however, can only tolerate a rotations up to 60 degrees. The TPS-RPM algorithm often fails to converge to a useful solution if rotation with any degree is allowed [2], so a parameter λ_2 is used to penalize a large rotation in the TPS-RPM algorithm. If λ_2 is set to zero, its performance deteriorates significantly, much worse than our approach at any level of rotation. Therefore, the default setting of λ_2 ($\lambda_2 = 0.01$) is used in this comparison experiment for the TPS-RPM algorithm.

The variance of all algorithms is large. Therefore, a statistical analysis must be applied to ascertain whether the difference between these algorithms is significant. Mean and variance can fully characterize only a Gaussian distribution. Fig. 5.12a and b show the error histograms of the shape context algorithm and our approach. The histograms are generated on 100 trials of the fish shape under the deformation level of 0.05. The distributions differ significantly from a Gaussian distribution. Some challenging samples deteriorate the performance and increase the variance, and

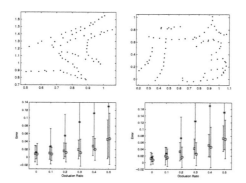

Figure 5.10: Comparison of our results (o) with the TPS-RPM (∗) and shape context (□) algorithms under occlusion. Left column: the fish shape. Right column: the Chinese character shape. Top row: synthesized samples. Bottom row: mean and variance of errors.

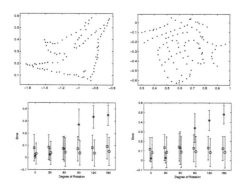

Figure 5.11: Comparison of our results (o) with the TPS-RPM (∗) and shape context (□) algorithms under rotation. Left column: the fish shape. Right column: the Chinese character shape. Top row: synthesized samples. Bottom row: mean and variance of errors.

the performance of two algorithms on the same sample is not independent. Fig. 5.12c shows the histogram of paired differences between two algorithms (the error of the shape context algorithm minus that of our approach). The two algorithms have the same performance for about one third of the test samples, and our approach outperforms the shape context algorithm on most of the remaining samples.

Since the distribution of errors is not Gaussian, we use the Wilcoxon paired signed rank test, which is powerful and distribution free [151]. In the Wilcoxon test, paired differences are formed, and the absolute values are ranked. Where ties occur, the average of the corresponding ranks is used. If the difference between two measures is zero, this sample is excluded from the analysis. The sum of the ranks with a positive sign and the sum of the ranks with a negative sign are calculated. The test statistic is the smaller of these two sums. Table 5.1 shows the statistical analysis (with two-sided significance level of 0.01) of the performance of our approach compared with two other algorithms. Here, $+ (-)$ means the improvement (deterioration) of our approach is statistically significant compared with the other algorithm, and $=$ means two algorithms do not differ significant. The statistical test verifies that the improvement of our approach on most data sets is significant.

5.7 Conclusions and Future Work

In this chapter, we have introduced the notion of a neighborhood structure for the general point matching problem. We formulated point matching as an optimization problem to preserve local neighborhood structures during matching. Extensive

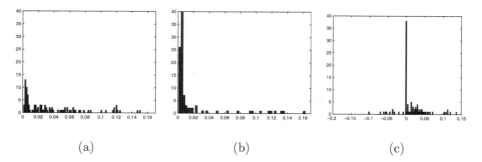

Figure 5.12: Histogram of errors. (a) The shape context algorithm. (b) Our approach. (c) Paired differences between two methods (the error of the shape context algorithm minus that of our approach).

Table 5.1: Wilcoxon paired signed rank test. $+$, $-$, and $=$ mean the former algorithm is better, worse, or no difference than the latter, respectively.

	Fish		Chinese character	
	Ours vs. shape context	Ours vs. TPS-RPM	Ours vs. shape context	Ours vs. TPS-RPM
Deformation	= = + + +	= = + + +	+ + + + +	+ + + + +
Noise	= + + + + +	+ = + + + +	+ + + + = +	+ + + + + +
Outlier	+ + + + +	+ + + = =	+ + + + +	+ = = − −
Occlusion	= + + + + =	+ + + + + =	+ + + + + +	+ + + + + +
Rotation	+ + + + + +	= + + = = =	− − + + + +	= = + + + +

137

experiments were presented to demonstrate the robustness of our approach. Compared with the other two state-of-the-art algorithms, our approach performs as well or better under nonrigid deformation, noise in point locations, outliers, occlusion, and rotation.

Large outlier or occlusion ratios (especially if the occlusion breaks a shape into several isolated parts) can significantly change the local neighborhood structure. A combination of different sources of degradation, such as large rotation, noise, and occlusion, also presents a challenge, which should be addressed in our future research. In this work, the relaxation labeling method is used to solve the constrained optimization problem. Only converging to a local optimum, it is by no means the best approach. We are testing other optimization methods such as simulated annealing, genetic algorithms, and graduated nonconvexity methods. Our graph matching formulation is applicable for both 2-D and 3-D shapes. Using the shape context distance for initialization, we only demonstrate it on 2-D shapes, since the original shape context is defined only for 2-D point sets. We will test the effectiveness of our approach for 3-D shape matching by extending the shape context to 3-D point sets.

A reference C++ implementation of our approach is available under the terms of the GNU General Public License (GPL) at `http://www.umiacs.umd.edu/~zhengyf/PointMatching.htm`.

Acknowledgments

The TPS-RPM algorithm used in the comparison experiments is based on the MATLAB® implementation by Dr. Haili Chui and Prof. Anand Rangarajan. The source code of the shape context algorithm came from Prof. Serge Belongie and Prof. Jitendra Malik. The synthesized data sets used in Section 5.6.2 were generated by Chui and Rangarajan, and provided to us by Belongie. We want to thank them for their help for making the comparison experiments much easier.

Appendix

Sinkhorn showed that iterated alternative row and column normalization will convert an $N \times N$ matrix with positive elements to a doubly stochastic matrix [143]. In our relaxation labeling approach, we perform iterated alternative row and column normalization (except the last row and column) to a non-square $K \times N$ ($K \neq N$) matrix A. This appendix demonstrates this approach is mathematically sound: the process will converge to a unique matrix T_A, such that each row and column of T_A sums one (except the last row and column). The proof in this appendix adheres to Sinkhorn's approach. In [143], several important steps are skipped and a few typographical errors exist. In this appendix, more cases are discussed to generalize Sinkhorn's conclusion. First, we give a formal definition of our generalized doubly stochastic matrix..

DEFINITION 1. A $K \times N$ matrix A is called a generalized doubly stochastic matrix if

$$\sum_{i=1}^{K} a_{ij} = 1 \quad \text{for } j = 1, 2, \ldots, N-1 \tag{5.28}$$

$$\sum_{j=1}^{N} a_{ij} = 1 \quad \text{for } i = 1, 2, \ldots, K-1 \tag{5.29}$$

The operation of row normalization can be represented as a left multiplication of A with a diagonal matrix, and the operation of column normalization can be represented as a right multiplication of A with another diagonal matrix. Multiple row (column) normalization matrices can be combined as D_1 (D_2). Therefore, the overall iterated row and column normalization can be represented as $T_A = D_1 A D_2$. The following theorem establishes the uniqueness of such a representation.

Theorem 5.1 *To a given strictly positive $K \times N$ matrix A there corresponds exactly one generalized doubly stochastic matrix T_A which can be expressed in the form $T_A = D_1 A D_2$ where D_1 and D_2 are diagonal matrices with positive diagonals. $D_1 = diag\{d_{11}, d_{12}, \ldots, d_{1,K-1}, 1\}$ and $D_2 = diag\{d_{21}, d_{22}, \ldots, d_{2,N-1}, 1\}$. The matrices D_1 and D_2 are unique.*

Proof: Suppose there exist two different pairs of diagonal matrices $D1$, D_2 and C_1, C_2 such that $P = C_1 A C_2$ and $Q = D_1 A D_2$ are both generalized doubly stochastic. Then, we can write Q as $Q = D_1 C_1^{-1} P C_2^{-1} D_2$. Let $E = D_1 C_1^{-1}$ and $F = C_2^{-1} D_2$, then $Q = EPF$. Suppose $E = diag\{e_1, e_2, \ldots, e_{K-1}, 1\}$ and $F =$

$\operatorname{diag}\{f_1, f_2, \ldots, f_{N-1}, 1\}$, Q can be expanded as

$$Q = \begin{bmatrix} e_1 f_1 P_{11} & e_1 f_2 P_{12} & \cdots & e_1 f_{N-1} P_{1,N-1} & e_1 P_{1N} \\ e_2 f_1 P_{21} & e_2 f_2 P_{22} & \cdots & e_2 f_{N-1} P_{2,N-1} & e_2 P_{2N} \\ \vdots & \vdots & \vdots & \vdots & \vdots \\ e_{K-1} f_1 P_{K-1,1} & e_{K-1} f_2 P_{K-1,2} & \cdots & e_{K-1} f_{N-1} P_{K-1,N-1} & e_{K-1} P_{K-1,N} \\ f_1 P_{K1} & f_2 P_{K2} & \cdots & f_{N-1} P_{K,N-1} & P_{K,N} \end{bmatrix}$$

$$(5.30)$$

The summation of the ith row of Q equals 1, for $1 \leq i \leq K - 1$.

$$e_i(f_1 P_{i1} + f_2 P_{i2} + \ldots + f_{N-1} P_{i,N-1} + P_{iN}) = 1 \qquad (5.31)$$

Since $\sum_{j=1}^{N} P_{ij} = 1$ and $P_{ij} > 0$, $1/e_i$ is a convex combination of $\{f_j, 1\}$. Therefore,

$$\min_j \{1, f_j\} \leq \frac{1}{e_i} \leq \max_j \{1, f_j\} \quad \text{for } i = 1, 2, \ldots, K - 1 \qquad (5.32)$$

Similarly, we can get

$$\min_i \{1, e_i\} \leq \frac{1}{f_j} \leq \max_i \{1, e_i\} \quad \text{for } j = 1, 2, \ldots, N - 1 \qquad (5.33)$$

There are three cases: 1) $\max_i e_i \leq 1$; 2) $\min_i e_i \geq 1$; and 3) $\min_i e_i \leq 1 \leq \max_i e_i$. Let's discuss the first case that $\max_i e_i \leq 1$. Using the second inequality in Eq. (5.33), we get $f_j \geq 1$. Then second inequality in Eq. (5.32) becomes $1 \leq e_i \max_j f_j$. It follows that

$$1 \leq \min_i e_i \max_j f_j \qquad (5.34)$$

Similarly, the first inequality in Eq. (5.33) becomes $f_j \min_i e_i \leq 1$. Therefore,

$$\min_i e_i \max_j f_j \leq 1. \qquad (5.35)$$

141

Combining the above two inequalities, we get

$$\min_i e_i \max_j f_j = 1 \qquad (5.36)$$

Let consider the summation of the row of Q corresponding to the minimum e_i. Suppose $e_1 = \min_i e_i$

$$
\begin{aligned}
1 &= e_1(f_1 P_{11} + f_2 P_{12} + \ldots + f_{N-1} P_{1,N-1} + P_{1N}) \\
&\leq e_1[\max_j f_j(P_{11} + P_{12} + \ldots + P_{1,N-1}) + P_{1N}] \\
&\leq e_1 \max_j f_j(P_{11} + P_{12} + \ldots + P_{1N}) \\
&\leq e_1 \max_j f_j \\
&= 1 \qquad\qquad\qquad\qquad\qquad\qquad\qquad (5.37)
\end{aligned}
$$

The equality holds if and only if $f_1 = f_2 = \cdots = f_{N-1} = 1$. And considering the column with the maximum f_j, we get $e_1 = e_2 = \cdots = e_{K-1} = 1$.

For the second case, $\min_i e_i \geq 1$, it is easy to verify that

$$\max_i e_i \min_j f_j = 1 \qquad (5.38)$$

And for the last case, $\min_i e_i \leq 1 \leq \max_i e_i$, we can get both equalities (5.36) and (5.38). Following similar arguments, we can show that the equalities $f_1 = f_2 = \cdots = f_{N-1} = 1$ and $e_1 = e_2 = \cdots = e_{K-1} = 1$ hold for all cases. It follows that $D_1 = C_1$, $D_2 = C_2$, and $P = Q$. That means such factorization is unique, and the resulted generalized doubly stochastic matrix is unique too.

Theorem 5.2 *The iterative process of alternately normalizing the rows and columns (except the last row and column) of a strictly positive $K \times N$ matrix is convergent to a strictly positive generalized doubly stochastic matrix.*

Proof: The iteration produces a sequence of positive matrices which alternately have row (except the last row) and column (except the last column) sums one. We will show that the two subsequences which are composed respectively of the matrices with row sums one and the matrices with column sums one each converge to a positive generalized doubly stochastic limit of the form $D_1 A D_2$. The uniqueness part of Theorem 5.1 guarantees two limits are the same. In the following, we only show the convergence of the subsequence of the matrices with column sums one. The convergence of the other subsequence is easy to show following similar arguments.

Let $\{A_n\} = \{(a_{nij})\}$ be the sequence with column sums one (except the last column), and A_n have row sums $\lambda_{n1}, \lambda_{n2}, \ldots, \lambda_{n,K-1}$. After row normalization, we calculate the column sums δ_{nj} (for $1 \leq j \leq N-1$)

$$\delta_{nj} = \sum_{i=1}^{K-1} a_{nij}/\lambda_{nj} + a_{nKj} \tag{5.39}$$

Since $\sum_{i=1}^{K} a_{nij} = 1$, δ_{nj} is a convex combination of $\{1/\lambda_{nj}, 1\}$. It follows

$$\frac{1}{\max\{1, \lambda_n(M)\}} \leq \delta_{nj} \leq \frac{1}{\min\{1, \lambda_n(m)\}} \quad \text{for } j = 1, 2, \ldots, N-1 \tag{5.40}$$

where the m and M respectively label minimal and maximal quantities relative to a given A_n. Similarly, since $\lambda_{n+1,i}$ of matrix A_{n+1} is a convex combination of $\{1/\delta_{nj}, 1\}$, it follows that

$$\frac{1}{\max\{1, \delta_n(M)\}} \leq \lambda_{n+1,i} \leq \frac{1}{\min\{1, \delta_n(m)\}} \quad \text{for } i = 1, 2, \ldots, K-1 \tag{5.41}$$

There are three cases: 1) $\lambda_n(m) \geq 1$; 2) $\lambda_n(M) \leq 1$; and 3) $\lambda_n(m) \leq 1 \leq \lambda_n(M)$. For the first case $\lambda_n(m) \geq 1$, from Eq. (5.40) we get $1/\lambda_n(M) \leq \delta_{nj} \leq 1$.

Using Eq. (5.41), we get

$$1 \leq \lambda_{n+1,i} \leq \lambda_n(M) \tag{5.42}$$

Therefore,

case 1: $\quad \lambda_n(m) \geq 1 \quad \Rightarrow \quad 1 \leq \lambda_{n+1}(m) \quad$ and $\quad 1 \leq \lambda_{n+1}(M) \leq \lambda_n(M)$ (5.43)

Similarly

case 2: $\qquad \lambda_n(M) \leq 1 \quad \Rightarrow \quad \lambda_{n+1}(M) \leq 1 \quad$ and $\quad \lambda_n(m) \leq \lambda_{n+1}(m)$ (5.44)

case 3: $\quad \lambda_n(m) \leq 1 \leq \lambda_n(M) \quad \Rightarrow \quad \lambda_n(m) \leq \lambda_{n+1}(m) \leq 1 \leq \lambda_{n+1}(M) \leq \lambda_n(M)$ (5.45)

In the following, we want to show that for case 1 and 3, $\lambda_n(M)$ left converges to 1 (from a value larger than 1); and for case 2 and 3, $\lambda_n(m)$ right converges to 1 (from a value smaller than 1). If the convergence holds, using Eq. (5.40), it follows that δ_{nj} converges to 1 too. Therefore, the sequence of matrices A_n converges to a generalized doubly stochastic matrix.

Let a_n be the minimal element of A_n (excluding the last row and column), we want to show that $a_n > 0$ for all n. Starting from $A_1 = \{a_{1ij}\}$, we can combine all row normalizations of row i ($i < K$) up to nth iteration as $x_{ni} = [\lambda_{1i}\lambda_{2i} \cdots \lambda_{ni}]^{-1}$. For the last row $x_{nK} = 1$. All column normalization of column j ($j < n$) up to nth iteration is combined as $y_{nj} = [\delta_{1j}\delta_{2j} \cdots \delta_{nj}]^{-1}$. For the last column $y_{nN} = 1$. Since summation of column j of A_n equals one, $\sum_{i=1}^{K} x_{ni}a_{ij}y_{nj} = 1$, for $j = 1, 2, \ldots, N-1$, we get

$$y_{nj} = \frac{1}{\sum_i a_{1ij}x_{ni}} \leq \frac{1}{a_{1ij}x_{ni}} \leq \frac{1}{a_1 x_{ni}} \tag{5.46}$$

144

In particular $y_{nj} \le 1/[a_1 x_n(M)]$. Since

$$\sum_{j=1}^{N} x_{ni} a_{1ij} y_{nj} = \lambda_{n+1,i} \qquad (5.47)$$

As we can see from (5.43), (5.44) and (5.45), for all three cases, $\lambda_{n+1,i}$ is bounded away from 0. Let $\lambda_{n+1,i} \ge \lambda$, it follows that

$$x_{ni} \ge \frac{\lambda}{\sum_j a_{1ij} y_{nj}} \ge a_1 \lambda x_n(M)/N. \qquad (5.48)$$

The last inequality is derived from the fact that $a_{1ij} \le 1$. Also $y_{nj} = 1/\sum_i a_{1ij} x_{ni} \ge 1/[N x_n(M)]$ and we see that $a_{n+1,i,j} = x_{ni} a_{1ij} y_{nj} \ge a_1 \lambda/N^2 = a > 0$. Therefore, $a_n > 0$ for all n.

For case 1 and 3, we want to show that $\lambda_n(M)$ right converge to 1. It is clear that $\lambda_n(M) \to 1 + c$ where $c \ge 0$. For convenience set $\lambda_n(M) = 1 + c_n$.

$$
\begin{aligned}
\delta_{nj} &= \sum_{i=1}^{K-1} \frac{a_{nij}}{\lambda_{ni}} + a_{nKj} = \sum_{i:\lambda_{ni} \le 1} \frac{a_{nij}}{\lambda_{ni}} + \sum_{i:\lambda_{ni} > 1} \frac{a_{nij}}{\lambda_{ni}} + a_{nKj} \\
&\ge \sum_{i:\lambda_{ni} \le 1} a_{nij} + \frac{1}{1+c_n} \sum_{i:\lambda_{ni} > 1} a_{nij} + \frac{1}{1+c_n} a_{nKj} = \frac{\sum_{i=1}^{K} a_{nij} + c_n \sum_{i:\lambda_{ni} > 1} a_{nij}}{1+c_n}
\end{aligned} \qquad (5.49)
$$

Using the fact that $\sum_i a_{nij} = 1$,

$$\delta_{nj} \ge \frac{1 + c_n \sum_{i:\lambda_{ni} > 1} a_{nij}}{1+c_n} \ge \frac{1 + c_n a_n}{1+c_n} \qquad (5.50)$$

It follows that

$$\lambda_{n+1,i} = \sum_{j=1}^{N-1} \frac{a_{nij}}{\lambda_{ni} \delta_{nj}} + \frac{a_{niN}}{\lambda_{ni}} \le \frac{1+c_n}{1+c_n a_n} \sum_{j=1}^{N-1} \frac{a_{nij}}{\lambda_{ni}} + \frac{a_{niN}}{\lambda_{ni}} \qquad (5.51)$$

Since $0 < a_n < 1$, therefore $(1+c_n)/(1+c_n a_n) > 1$, thus

$$\lambda_{n+1,i} \le \frac{1+c_n}{1+c_n a_n} \left(\sum_{j=1}^{N-1} \frac{a_{nij}}{\lambda_{ni}} + \frac{a_{nNj}}{\lambda_{ni}} \right) \qquad (5.52)$$

145

Because $\sum_{j=1}^{N} a_{nij}/\lambda_{ni} = 1$ (the row summation after row normalization), therefore,

$$\lambda_{n+1,i} \leq \frac{1+c_n}{1+c_n a_n} < \frac{1+c_n}{1+c_n a} \qquad (5.53)$$

The above inequality holds for all i, particularly,

$$1 + c \leq \lambda_{n+1}(M) < \frac{1+c_n}{1+c_n a} \qquad (5.54)$$

Since $c_n \to c$, the above condition holds if and only if $c = 0$. Therefore $\lambda_n(M) \to 1$.

For case 2 and 3, we need to show that $\lambda_n(m)$ left converge to 1. Let $\lambda_n(m) \to 1 - d$ where $d \geq 0$, and $\lambda_n(m) = 1 - d_n$, then

$$\delta_{nj} = \sum_{i:\lambda_{ni}\leq 1} \frac{a_{nij}}{\lambda_{ni}} + \sum_{i:\lambda_{ni}>1} \frac{a_{nij}}{\lambda_{ni}} + a_{nMj} \leq \frac{1}{1-d_n} \sum_{i:\lambda_{ni}\leq 1} a_{nij} + \sum_{i:\lambda_{ni}>1} a_{nij} + a_{nMj} = \frac{1-d_n a_n}{1-d_n}$$

$$(5.55)$$

And

$$1 - d \geq \lambda_{n+1}(m) \geq \frac{1-d_n}{1-d_n a_n} > \frac{1-d_n}{1-d_n a} \qquad (5.56)$$

Since $d_n \to d$, the above condition holds if and only if $d = 0$. It follows $\lambda_n(m) \to 1$.

This completes the proof.

Chapter 6

Handwriting Synthesis

6.1 Introduction

A statistical pattern recognition system depends heavily on the size and quality of the training set. Although preparing samples of machine printed text is easy, doing so is expensive for handwriting. Synthesized data can be used as a supplement. In the previous work, many handwriting synthesis approaches have been proposed to simulate the writing style of a person [152], or enlarge the training set for a recognition system [108, 3]. They can be roughly categorized as perturbation-based, model-based, or example-based. Perturbation-based methods need only one handwriting sample. New samples are generated by assigning random parameters to a deformation model, which is then used to deform the sample [108, 2]. However, without considering the deformation characteristics of handwriting, unrealistic samples may be generated. Instead, model-based approaches learn the deformation of handwriting and explicitly describe it as a distribution (the distribution is often called a model) [152, 153]. After learning, handwriting synthesis is the process of drawing new samples from the distribution. Although theoretically founded, model-based approaches have some drawbacks in real applications: handwriting models are often complex, and many samples are demanded for model training. Example-based approaches use two handwriting samples and generate new samples with shapes similar to both inputs [3]. Compared with model-based approaches, fewer samples are needed because this approach does not need to learn the distribution of deformation. Both model-based and example-based approaches need to perform handwriting matching, which is a challenge because handwriting is a nonrigid shape.

6.2 Our Approach

The key problem of handwriting synthesis involves generating samples that look natural. Otherwise, arbitrarily synthesized samples cannot improve (if not deteriorate) the performance of the system trained on them. Although handwriting samples vary greatly in respect to size, rotation, and stroke width, shape is generally used to categorize them into different classes. Since nonrigid deformation of handwriting is large, we argue that a synthesis algorithm should learn the shape deformation characteristics from real handwriting samples. It is reasonable to assume that the shape space of handwriting with the same content (e.g., the handwriting samples of the letter 'a') is continuous. For characters with several different writing glyphs, such as number '7,' we may need to do clustering analysis to segment the shape space into multiple continuous sub-spaces. Given two handwriting samples close in the shape space, the interpolation between them is likely to lie inside the shape space too (this is guaranteed if the shape space is convex). That means, given two actual and similar handwriting samples, it is reasonable to assume a person may write with a shape between them (i.e., with similar but less degree of deformation).

In this chapter, we propose an example-based handwriting synthesis approach using two training samples. We use our handwriting matching algorithm to establish the correspondence between two handwriting samples. After handwriting matching, we warp one sample toward the other using the TPS deformation model. By adjusting the regularization parameter λ of the TPS deformation model (Equation (6.1)), we can adjust the amount of non-rigid deformation.

$$H_f = \sum_{i=1}^{n}[z_i - f(x_i, y_i)]^2 + \lambda I_f \tag{6.1}$$

Please refer to Section 5.4 for a detailed description of the regularization. The regularization parameter λ is used in a different way compared to the shape matching in the previous chapter. Here, we use it to adjust the amount of non-rigid deformation. It has been shown that if $\lambda = \infty$, the deformation is the affine transformation. With a smaller λ, the interpolated shape is closer to the target shape. On shape matching, however, it is used to reduce the effect of outliers in the match estimate.

Among all previous work, the algorithm proposed by Mori et al. [3] is the most similar to our approach. They use the well-known iterated closed point (ICP) algorithm [111] to get the displacement vector of each pixel. A new sample is generated by moving each pixel along its displacement vector. Compared with our approach, it has two drawbacks: (1) the ICP algorithm is not robust under nonrigid deformation [2], and (2) the displacement field is not continuous, so the synthesized sample may change the topology.

149

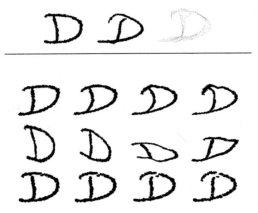

Figure 6.1: Handwriting synthesis. Two training samples and the point matching result are shown on the top row. Synthesized samples: second row for our approach, third row for [2], and forth row for [3].

6.3 Experiments

In this section, we apply our handwriting matching algorithm to handwriting synthesis and compare it with two other algorithms [2, 3]. The top row of Fig. 6.1 shows two handwriting samples and their point matching result. Under this match, we use TPS to deform the first sample to synthesize new samples, as shown in the second row. From left to right, λ takes the value of ∞, 10, 1, and 0.1, respectively. With the decrease of the regularization parameter λ, the synthesized sample is closer to the second real sample. The third row in Fig. 6.1 shows synthesized samples using a perturbation-based approach [2]. With only one training sample, the deformation of synthesized samples is random, and sometimes unnatural samples may be generated. The third row shows the synthesized samples using the approach of [3]. Suppose point T_i is matched to point $D_{f(i)}$, its displacement is $d_i = D_{f(i)} - T_i$. For

150

an un-matched point, the displacement takes the value of its closest matched point. A new sample is synthesized by moving a point along its displacement.

$$S_i = T_i + \rho d_i \qquad (6.2)$$

If $\rho = 0$, the synthesized sample is the original template; and if $\rho = 1$, all matched points will be moved to the target under any matching function. The last row of Fig. 6.1 shows several synthesized samples with ρ equals 0.2, 0.4, 0.6, and 0.8, respectively. As shown in the figure, both example-based approaches can learn shape deformation characteristics given good point matching results. The approach of [3], however, may change the topology (the synthesized handwriting is broken into several parts) due to the dis-continuity of the displacement field. This draw-back becomes obvious when point matching errors (which are un-avoidable in real applications) are present. The second and third rows of Fig. 6.2 show synthesized samples using our approach and [3]. The topology of samples synthesized by our approach is preserved, even under substantial matching errors. Using the approach of [3], unrealistic samples are generated.

More examples on handwriting synthesis are shown in Fig. 6.3. The ordering of samples is the same as Fig. 6.1. Samples with different slants (within the range of the slant between two training samples) are generated using our approach.

6.4   Discussion and Future Work

In this chapter, we applied our handwriting matching algorithm to synthesize new training samples for handwriting recognition. Our approach automatically learns

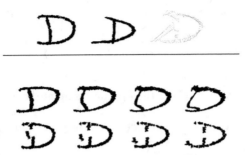

Figure 6.2: Handwriting synthesis with point matching errors presented. Two training samples and the point matching result are shown on the top row. Synthesized samples: second row for our approach, third row for [3].

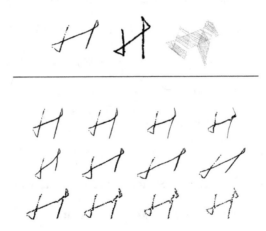

Figure 6.3: More examples on handwriting synthesis. Two training samples and the point matching result are shown on the top row. Synthesized samples: second row for our approach, third row for [2], and forth row for [3].

shape deformation characteristics from training samples, generating more visually realistic samples. Although it has been shown in several independent experiments that synthesized handwriting samples can improve a recognition system trained on them [3, 108], we will perform more experiments to verify it in the future. The limitation of our approach is that we assume the same character is written with similar shape though with some degree of distortion. Due to difference in geological location, education, and culture, people may write the same character with significantly different shapes. We may need to perform clustering analysis to separate the training samples into several clusters. Each cluster is more homogeneous, so our approach can be applied.

Chapter 7

Handwriting Retrieval

7.1 Introduction

Handwriting not only conveys information by its content (what one writes), but also contains unique characteristics of the writer (how one writes). People have produced huge amounts of handwritten documents in history and continue to do so. With the coming of the electronic document era, the traditional library faces the challenge of how to make an enormous amount of handwritten historical documents accessible electronically. Scanning documents into a computer is easy. However, without a searchable index, the scanned images have little use. In one solution, optical character recognition (OCR) can convert handwriting to text, then the traditional text-based retrieval techniques can perform their tasks. However, the state-of-the-art techniques for handwriting recognition are error prone. No reliable product is available to recognize unconstrained handwriting, except for handwritten digits [6]. Recognition errors will significantly deteriorate the performance of the traditional text-based techniques. Direct retrieval based on image, without recognition, may achieve better results [154]. Another application of handwriting retrieval is searching a specified signature in a set of documents. Handwriting recognition has little use in this application because the signature's character sequence has little or no inherent importance. Instead, its unique characteristics are of more concern. In these applications, the direct retrieval of handwriting in document images is desired.

The retrieval of character sequences is also called *keyword spotting* in the literature [155]. Two similar keyword spotting approaches based on the hidden Markov models (HMM) were proposed in [155] and [156] for printed text in degraded document images, where the results of OCR were not reliable. The training of HMMs needs a large set of labeled samples, but only one or several handwriting instances may be available for query. Several handwriting retrieval techniques, such as the dynamic time warping based approach [154] and the corner feature based approach [157], have been proposed. However, these techniques are not robust under non-rigid deformation of handwriting.

Alternatively, we can consider handwriting as a shape. Shape presentation, analysis, matching, and recognition is an active area in computer vision and widely studied [109, 110]. Recently, the shape context method proposed by Belongie et al. [26] has achieved success in many shape recognition and retrieval applications. A shape is represented as a set of points in this approach. A shape context is assigned to each point, which describes the relative distribution of the other points in the shape. After defining the similarity between two points based on their shape contexts, the Hungarian algorithm [126] searches for the optimal match between the two point sets. Similarity measures, such as the shape context distance and the thin plate spline (TPS) bending energy, are calculated with the current estimate of the correspondence and used for shape recognition or retrieval. Ling and Jacobs [158] improved the original shape context method by replacing the Euclidean distance with the inner-distance, which is invariant for articulated shapes, in the calculation of shape context. Though the discrimination power is improved by using the inner-distance, the method is sensitive to structure change of a shape. As shown in Fig. 7.3, a handwriting stroke is often broken. Two strokes may touch each other in one sample, but well separated in another sample. Structure change is also a challenge for shock-graph based approaches [159, 160], where the connection of two points by a stroke is utilized. On the contrary, by treating a shape as a set of isolated points, the original shape context method is more robust under structure change. In many shape retrieval approaches [161, 162, 158], a shape is often represented as a sequence of ordered points sampled from its contour. This may apply for the retrieval of on-line handwriting samples collected on a PDA or TablePC. However, for an off-line handwriting sample, it is difficult to recover the temporal information [163]. Therefore, we cannot order points to form a 1-D sequence reliably due to structure change of handwriting.

In this chapter, we test several variations of the original shape context method for handwriting retrieval. The shape context method can be decomposed as two steps: shape matching and shape retrieval. Each step has several alternative technical solutions, which are tested in our experiments. To avoid confusion, we call these two steps of the original method the shape context matching algorithm and the shape context retrieval algorithm, respectively. As discussed in the previous chapter, the shape matching part of the shape context algorithm is not robust. Two neighboring points in one shape may be matched to two points far apart in the other shape. We proposed a new shape matching algorithm in the previous chapter to preserve local neighborhood structures during matching. Experiments show our new approach achieves much better shape matching results. In this chapter, we replace the original shape matching method with the proposed one and test it for handwriting retrieval. Experiments show that a slight improvement is achieved. We propose a few new similarity measurements based on the point matching results to further improve the retrieval accuracy. The performance with a single query is limited due to noise and the large variations of handwriting. We propose a method to combine retrieval results with multiple queries.

7.2 Our Handwriting Retrieval Method

In the original shape context method, a set of uniformly sampled points on the contour is used to represent a shape, as contour is a general representation and a shape can be recovered from its contour. However, a contour is affected by the stroke width, which depends on the writing tools and scanning parameter settings. Instead, we use a skeleton, which is widely used in handwriting recognition, to represent a handwriting sample. There are two advantages of using a skeleton: 1) it is robust under the variation in stroke width; 2) less points are demanded to represent handwriting. Suppose each shape is represented with N points, the computation time is in the order of $O(N^3)$ for shape matching and $O(N^2)$ for similarity calculation. With a small number of points, we can significantly increase the retrieval speed. Fig. 7.1a shows the shape representation with 300 points sampled from the contour, and Fig. 7.1b shows the representation with 200 points from the skeleton. We can see that 200 points from the skeleton is as good as 300 points from the

(a)   (b)

Figure 7.1: Handwriting representation: contour vs. skeleton. (a) Representation with 300 points sampled from the contour. (b) Representation with 200 points sampled from the skeleton.

contour to represent a handwriting sample.

After point matching, two similarity measures were proposed in the original shape context method [26] for shape recognition and retrieval. One is based on the shape context distance ($\mathcal{D}_{sc}$), and the other is the TPS bending energy ($\mathcal{D}_{be}$). Suppose there are M points in the template shape T and N points in the deformed shape D. The shape context distance between shapes T and D is the symmetric sum of shape context matching costs over best matching points,

$$\mathcal{D}_{sc} = \frac{1}{M} \sum_{t \in T} \arg\min_{d \in D} C(\mathcal{T}(t), d) + \frac{1}{N} \sum_{d \in D} \arg\min_{t \in T} C(\mathcal{T}(t), d), \qquad (7.1)$$

where $\mathcal{T}(.)$ denotes the estimated TPS shape transformation; $C(.,.)$ is the shape context matching cost between two points. Consider two points, t in shape T, and d in the other shape D. Their shape contexts are $h_t(k)$ and $h_d(k)$, for $k = 1, 2, \ldots, K$, respectively. $C(t, d)$ is defined as

$$C(t, d) = \frac{1}{2} \sum_{k=1}^{K} \frac{[h_t(k) - h_d(k)]^2}{h_t(k) + h_d(k)}. \qquad (7.2)$$

The TPS bending energy represents the amount of deformation between two shapes and can be used as a shape similarity measure. As discussed in the previous chapter, two TPS models are used for 2-D coordinate transformation. $\mathcal{D}_{be}$ is defined as the sum of the bending energy of two TPS models.

Besides $\mathcal{D}_{sc}$ and $\mathcal{D}_{be}$ proposed in the original shape context method, we can define more similarity measures to improve the retrieval accuracy. The TPS bending energy only measures the energy of deformation beyond an affine transformation. In other words, the bending energy is zero for an affine transformation. An affine transformation can be decomposed as translation, rotation, and anisotropic scales. According to Kendell's shape theory [164][145, p. 1], shape is, "all the geometrical information that remains when location, scale and rotational effects are filtered out from an object." So, shape is invariant up to the similarity transformation (translation, rotation, and isotropic scale). The amount of anisotropic scales is a good measure of the similarity between two shapes. [1] Suppose the scales of the x and y coordinates are S_x and S_y, respectively. We can estimate the scales from the affine transformation matrix based on the singular value decomposition. A 2-D affine transformation can be represented as a 2×2 matrix A and a 2×1 translation vector T

$$\begin{pmatrix} u \\ v \end{pmatrix} = A \begin{pmatrix} x \\ y \end{pmatrix} + T. \tag{7.3}$$

---

[1] This similarity measure was implemented in the Matlab package of the shape context method distributed on-line. But its effectiveness in shape recognition and retrieval has not been tested and documented in the paper [26].

S_x and S_y are the singular values of matrix A. The distance $\mathcal{D}_{af}$ is defined as

$$\mathcal{D}_{af} = \log\left(\frac{\max(S_x, S_y)}{\min(S_x, S_y)}\right). \tag{7.4}$$

If the scales are isotropic, $S_x = S_y$, then $\mathcal{D}_{af} = 0$.

We propose another distance measure, $\mathcal{D}_{re}$, based on the registration residual errors. Suppose point t_i in shape T is matched to point $d_{f(i)}$ in shape D. $\mathcal{D}_{re}$ is defined as

$$\mathcal{D}_{re} = \frac{\sum_{i:f(i)\neq nil} \|\mathcal{T}(t_i) - d_{f(i)}\|}{\sum_{i:f(i)\neq nil} 1}, \tag{7.5}$$

where $\|.\|$ is the Euclidean distance. In calculating $\mathcal{D}_{re}$, we remove those points matching to a dummy point nil.

Our matching algorithm will automatically reject some points as outliers. If two shapes are similar, the number of rejected points will be small, so the number of outliers offers a measure of the similarity between two shapes. The distance measure $\mathcal{D}_{ou}$ is defined as the number of outliers rejected during matching.

In total, we have five similarity measures, which have different scales. The overall similarity measure $\mathcal{D}_{all}$ is defined as the weighted Euclidean distance of all distance measures. Given a query, we calculate its distance measures to all samples in the database. We, then, can calculate the variance of each measure. The overall similarity measure $\mathcal{D}_{all}$ is define as

$$\mathcal{D}_{all} = \frac{\mathcal{D}_{sc}}{Var(\mathcal{D}_{sc})} + \frac{\mathcal{D}_{be}}{Var(\mathcal{D}_{be})} + \frac{\mathcal{D}_{af}}{Var(\mathcal{D}_{af})} + \frac{\mathcal{D}_{re}}{Var(\mathcal{D}_{re})} + \frac{\mathcal{D}_{ou}}{Var(\mathcal{D}_{ou})}. \tag{7.6}$$

Since the weights are calculated on-line, different queries may have different combination weights. In general, we do not have many samples to train the weights for a retrieval task, and the above scheme is by no means optimal. Commonly, user's feedback is often used to adjust the weights in the retrieval community. This process is called relevance feedback, which may significantly increase the retrieval accuracy after a few iterations.

The retrieval performance with a single query instance is limited due to the noise and the large variations of handwriting (as shown in Figs. 7.2b and 7.3). The performance of a retrieval depends heavily on the sample used for query. It is often possible to have a couple of handwritten samples from the same person available for retrieval. For example, we can ask the person to write a sample several times, or we can select the correctly returned samples from the first-round retrieval and use them with the original query instance in the second-round retrieval. Suppose, multiple instances from the same class $q_1, q_2, \ldots, q_k$ are used as queries. The corresponding distances of a handwritten sample in the collection to these queries are $d_1, d_2, \ldots, d_k$. We combine them into a final distance

$$d = f(d_1, d_2, \ldots, d_k). \tag{7.7}$$

Using one instance for query, we can achieve a sorted list of all samples in the collection. Using multiple instances, multiple sorted lists of the same collection can be achieved. The problem of combining these sorted lists to get the final one is similar to the multiple classifier combination problem, which is an active research topic in pattern recognition. Many techniques have been proposed. Currently we use a simple technique: using the smallest distance as the combined distance measure

$$d = min\{d_1, d_2, \ldots, d_k\}. \tag{7.8}$$

161

7.3 Experiments

In this section, we first present our retrieval evaluation metrics. The effects of different shape representations, shape matching algorithms, and similarity measures are compared in the task of handwritten initial retrieval. We also present an experiment for logo retrieval.

7.3.1 Evaluation Metrics

Evaluation metrics for retrieval are well studied in the traditional information retrieval literature. Generally, two set of metrics, *precision* and *recall*, are widely used.

$$\text{Precision} = \frac{\# \text{ Returned relevant samples}}{\# \text{ Returned samples}} \tag{7.9}$$

$$\text{Recall} = \frac{\# \text{ Returned relevant samples}}{\# \text{ Relevant samples in the database}} \tag{7.10}$$

Region of characteristic (ROC) curves, indicating the relationship between precision and recall, can release the performance of a retrieval algorithm completely. Since comparing ROC curves of two algorithms is complicated, sometimes, a metric with a scalar value is preferred. *R-Precision* is such a metric. Suppose a total of M samples in the database, and among them N samples are relevant for a query. The retrieval algorithm will return a ranked list of all M samples. R-Precision is the precision of the first N retrieved samples in the list.

$$\text{R-Precision} = \frac{\# \text{ Relevant samples in top } N \text{ returned samples}}{N} \tag{7.11}$$

R-Precision is used for evaluation in the following experiments.

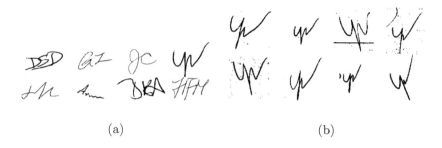

(a)   (b)

Figure 7.2: Some samples from the handwritten initial data set. (a) One sample from each person. (b) Several samples from one person.

7.3.2 Handwritten Initial Retrieval

To test the effectiveness of the proposed method for handwriting retrieval, we collected a small dataset of handwritten initials. We had eight persons with each of them writing 40 sets of initials. Fig. 7.2a shows one sample from each person, and 7.2b shows several samples from one person. As we can see, variations (besides rotation and scales) are large for handwriting, and noise, such as underlines, would make things worse.

In the first experiment, we evaluate the effect of different shape representations on retrieval accuracy. For a general shape, a contour is often use to represent a shape, then a number of points are uniformly sampled from the contour. Handwriting is composed of thin strokes, so skeleton representation may offer a better choice than contour. In this experiment, we use the original shape context algorithm for experiments: the original shape context matching method is used to search for the correspondence between two point sets; $\mathcal{D}_{sc}$ and $\mathcal{D}_{be}$ are used to measure the similarity between two shapes. The number of sampled points is fixed as 200. For

163

an exhaustively evaluation on the database, each sample should be selected once for query. Supposing M samples in the database, the number of shape comparisons is $M \times (M - 1)$. In our case, it is $320 \times 319 = 102,080$. Each shape comparison may take about one second. Due to the speed, we randomly select 10% of the samples as queries. It is sufficient to reveal the difference between different shape representations. To remove any bias, the selected sample is removed from the database for this query. The overall R-Precision is reported as the average of R-Precisions of all queries. The overall R-Precision of the contour based representation is 69.4%. For comparison, two skeleton extraction algorithms are implemented [165, 166]. The overall R-Precision increases slightly to 70.5% using Dyer and Rosefled's method [165], and to 71.4% using Zhang and Suen's method [166]. Since the latter skeleton extraction algorithm achieves the best performance, we use it for the following experiments.

In this experiment, we test the effectiveness of our shape matching method. We replace the matching method with our approach presented in the last chapter, and use the same similarity measures $\mathcal{D}_{sc}$ and $\mathcal{D}_{be}$. The same 10% randomly selected samples in the previous experiment are used as queries. The overall R-Precision using our shape matching approach is 72.3%. Out of our expectation, the improvement is slight compared to the original overall R-Precision of 71.4%. Detailed analysis shows that our shape matching approach decreases the average distances between shapes from the same category. This verifies that better shape matching results are achieved by our approach. However, our matching approach also decreases the average distances between shapes from different categories. The conclusion of this

164

experiment is that a better shape matching algorithm does not necessarily increase the discrimination capability of a measure, which is defined on the matching results.

In the following experiment, we compare the effectiveness of different similarity measures. Unlike the previous experiments, we make full use of the database. All samples are used in turn for query. Since our shape matching algorithm is much slower than the shape context matching algorithm, we use the original matching method. Table 7.1 shows the experimental results. The most powerful similarity measure is the TPS bending energy ($\mathcal{D}_{be}$), followed by the affine transformation based measure ($\mathcal{D}_{af}$). This experiment show that measures based on transformations (TPS for non-linear transformation and affine for linear transformation) are more robust than other features. The remaining measures can be ordered according to the discrimination power as the shape context distance ($\mathcal{D}_{sc}$), the residual error in registration ($\mathcal{D}_{re}$), and the outlier ratio ($\mathcal{D}_{ou}$). The outlier ratio feature is significantly less effective than the others, although it still provides some retrieval power. By combining $\mathcal{D}_{be}$ and $\mathcal{D}_{sc}$ as in the original shape context method, the overall R-Precision is 71.2%. If we combine all features, the overall R-Precision improves to 74.4%, demonstrating that different features are complementary. Fig. 7.3 shows one query example. Excluding the query itself from the database leaves 39 relevant samples. Among them, nine are ranked outside the top 39 returned samples. The figure also shows the missed samples and false alarms. Most false alarms rank low, just higher than the border line.

Table 7.2 shows the retrieval accuracy using multiple instances from the same person. We randomly select multiple instances from a person and remove them

Query Instance

Missed samples

(45)

(47)

(103)

(115)

(153)

(158)

(177)

(251)

(255)

False alarms

(28)

(30)

(31)

(32)

(34)

(35)

(36)

(37)

(39)

Figure 7.3: Handwritten initial retrieval. The first row is the instance used for query. The middle zone shows the missed samples, and the bottom zone shows the false alarms. The rank for each sample is shown under the corresponding image.

Table 7.1: Query with different similarity measures.

	$\mathcal{D}_{sc}$	$\mathcal{D}_{be}$	$\mathcal{D}_{af}$	$\mathcal{D}_{re}$	$\mathcal{D}_{ou}$	$\mathcal{D}_{sc}+\mathcal{D}_{be}$	$\mathcal{D}_{all}$
Overall R-Precision	56.1%	60.9%	57.3%	50.1%	35.5%	71.2%	74.4%

Table 7.2: Query with multiple handwritten initial samples from the same person.

Number of instances	One	Two	Three	Four
Overall R-Precision	74.4%	83.4%	86.2%	88.1%

from the database. Each instance is used to query the database. Eq. 7.8 is used to combine the overall distance measures $\mathcal{D}_{all}$ of multiple queries. As we can see, using multiple instances can significantly increase the performance. With more instances added, the overall R-Precision increases steadily. When four instances are used, 88.1% overall R-Precision is achieved.

7.3.3 Logo Retrieval

A logo is important to identify a document's source. Generally, logos can be seen as rigid shapes. However, some organizations make small adjustment to their logos periodically. Logos from different departments under the same organization (such as a university or the federal government) may contain a similar layout with a small variation to reflect their identity. In this experiment, we consider the task of finding all documents containing a specified logo given by a query. Since a logo is often embedded in a document, we need to segment it from other document components.

167

We use the automatic logo detection algorithm developed by Dr. Drayer at Fort Meade, Maryland. The tobacco data set is used for the experiment. In the data set, 546 documents contain visible logos, and the algorithm detected 350 of them. A logo is claimed to be detected if the major parts of the logo are detected, so in many cases the so-called *correct* detection is not perfect. A small part of a logo may be lost, and other components may be grouped into the detected logo. We use these 350 detected logos for the experiment of logo retrieval. There are 23 logo categories, and some may contain sub-categories (variations). The distribution of each category is not even. About half of the logos, 179 instances, belong to one category. We randomly selected one instance from this category as a query, as shown in the first row of Fig. 7.4. Since skeleton representation is sensitive to noise for a general shape, we use 200 points uniformly sampled from the contour to represent a logo. All similarity measures are used, and the overall distance measure $\mathcal{D}_{all}$ is used to rank the remaining 349 logos. Thirty-three of the relevant logos rank outside the top 178, resulting in the corresponding R-Precision of 81.5%. Fig. 7.4 shows several relevant logos that rank outside the top 178. Most missed logos occur because of the segmentation errors. In some cases, the approach may fail due to variations in the logo design, as shown in the last row of Fig. 7.4.

To fully evaluate the performance in logo retrieval, we use each instance in the collection for query. Table 7.3.3 shows the overall R-Precision using different similarity measures. The overall R-Precision is 67.9%. As demonstrated in the table, combining different similarity measures can significantly improve the retrieval accuracy.

 Query Instance

Missed samples due to segmentation errors

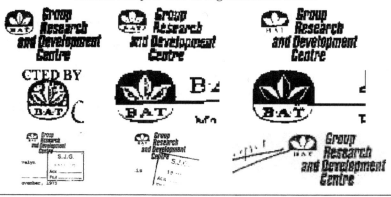

Missed samples due to variations in the logo design

Figure 7.4: Logo retrieval. The top row is the instance used for query. The following rows show the missed logos in the top returned samples.

Table 7.3: Overall R-Precision for logo retrieval with different similarity measures.

	$\mathcal{D}_{sc}$	$\mathcal{D}_{be}$	$\mathcal{D}_{af}$	$\mathcal{D}_{re}$	$\mathcal{D}_{ou}$	$\mathcal{D}_{all}$
Overall R-Precision	53.8%	57.8%	54.6%	48.3%	30.3%	67.9%

7.4 Summary and Future Work

Experiments show that shape context is effective for handwriting retrieval. Skeleton representation is more suitable for handwriting than contour and improves the retrieval performance. Our new shape matching method only slightly improves the retrieval accuracy, since it simultaneously decreases the distance measures between two shapes from both the same and different categories. Overall, the distinguishing power of a distance measure only improves slightly using our matching method. Adding more similarity measures will improve the retrieval accuracy significantly. A more effective way to improve the accuracy is to use multiple samples for query. When four instances are used for query, the overall R-Precision increases from 74.4% to 88.1% in our handwritten initial retrieval experiment.

Handwriting retrieval (non-rigid shape retrieval in general) is a difficult problem. In this chapter, we presented only our preliminary efforts on this topic. Much work remains for the future. We tried to combine our previous work on handwriting identification with the handwriting retrieval proposed in this chapter into a complete system. However, segmentation errors significantly deteriorated the overall retrieval performance. This is still an open problem in computer vision and needs further investigation. Future research may also look into the optimal combination scheme for different distance measures. Relevance feedback is a solution, but the effectiveness of this approach needs to be verified in the context of handwriting retrieval by experiments.

Chapter 8

Conclusions

8.1 Summary

Our work has centered around the ability to separate handwriting from other layers in noisy documents. Following are our key contributions.

1. Many handwritten documents have rule lines as a background pattern. The lines must be detected and removed before feeding the text to an optical character recognition (OCR) engine. Severely broken rule lines present a great challenge for existing line detection algorithms. We proposed an HMM model to incorporate the constraints among a set of parallel lines. Our method is fast and achieves both a high accuracy and a low false alarm rate. It has been tested on a real Arabic data set, and promising results were achieved. After line detection, line removal is performed by line width thresholding.

2. For handwriting identification in noisy documents, we proposed a classification based scheme. The input document is segmented at the word level. Several features, including Gabor filters, run-length histograms, crossing-count histograms, textures, and structural features are extracted for classification. The classification result is reasonable, with a few mis-classifications due to the overlapping of different classes in the feature space. Some other cues may re-

fine the classification results. For example, machine printed text, handwriting, and noise exhibit different patterns of geometric relationships. Printed words often form horizontal (or vertical) text lines, and noise blocks tend to overlap each other. The novelty of our approach involves using the Markov random field (MRF) to model the geometric relationship among neighboring blocks. Experiments show that MRF based post-processing is effective, where almost half of the mis-classifications are corrected after post-processing.

3. The identified handwriting may be further analyzed. In this work, we proposed a novel point pattern based handwriting matching technique and applied it for handwriting synthesis and retrieval. We studied handwriting matching in a broader context of nonrigid shape matching. For nonrigid shapes, most neighboring points cannot move independently under deformation due to physical constraints. Therefore, though the absolute distance between two points may change significantly, the neighborhood of a point is well preserved in general. Based on this observation, we formulate point matching as a graph matching problem. Each point is a node in the graph, and two nodes are connected by an edge if their Euclidean distance is less than a threshold. The optimal match between two graphs is the one that maximizes the number of matched edges. The shape context distance is used to initialize the graph matching, followed by relaxation labeling for refinement. Experiments demonstrate the effectiveness of our approach: it outperforms the shape context and TPS-RPM algorithms under nonrigid deformation and noise on a public data set.

4. The techniques proposed in this paper are not limited to the processing of handwriting documents. For example, our model-based line detection algorithm, proposed in Chapter 2, can extend straightforwardly for known form identification and registration. In Chapter 4, we separate different components into three layers, handwriting, machine printed text, and noise. By separating noise, the layer of machine printed text is much cleaner than the original noisy document, which facilitates further processing, such as zone segmentation and OCR. Our approach for handwriting matching, discussed in Chapter 5, is general enough to be applied to other point pattern based nonrigid shape matching applications.

8.2   Future Work

Though promising results have been achieved on several key issues in processing of handwriting documents, many possible extensions for further improvement exist.

1. In Chapter 4, how to extend our handwriting identification method to cursive scripts, such as Arabic, is under investigation. We found two observations used to discriminate handwriting from machine printed text for English documents do not work well for Arabic documents. (1) Handwriting is more cursive than machine printed text in English documents. However, machine printed Arabic text is cursive by nature. (2) People like to connect several neighboring characters during writing. However, machine printed Arabic characters are often connected too. Preliminary experiments on Arabic documents, using the same

feature set proposed in Chapter 4, resulted in a poor classification accuracy. One possible solution is to design new features for Arabic documents. Alternatively, we can perform handwriting/machine printed text discrimination at a higher level than word blocks, such as the text line level. Reliably extracting text lines from a heterogeneous and noisy document is a challenging problem itself. Our preliminary results using the level set method are promising, but more experiments are necessary. Another problem is that word level classification is still demanded in real applications because short text lines may contain only one or two words. How to combine the classification results at word and text line levels presents one direction for future research.

2. For nonrigid shape matching in Chapter 5, large outlier or occlusion ratios (especially if the occlusion breaks a shape into several isolated parts) can significantly change the local neighborhood structures. Combination of different sources of degradation, such as large rotation, noise, and occlusion, also presents a challenge, which should be addressed in future research. In this work, the relaxation labeling method is used to solve the constrained optimization problem. Converging only to a local optimum, it is by no means the best approach.

3. In Chapter 7, our experiments on handwriting retrieval show that a better shape matching method does not always result in a higher retrieval accuracy. Combining several robust similarity measures is more effective to improve the retrieval accuracy. How to get the optimal combination weights for different

measures is a topic under investigation. Techniques in the literature of traditional information retrieval, such as relevance feedback or clustering analysis, should be studied in the context of handwriting retrieval to test their effectiveness. Segmentation errors will significantly deteriorate the overall retrieval performance. This is still an open problem in computer vision and needs further investigation.

BIBLIOGRAPHY

[1] Y. Zheng, C. Liu, and X. Ding, "Form frame line detection with directional single-connected chain," in *Proc. Int'l Conf. Document Analysis and Recognition*, 2001, pp. 699–703.

[2] H. Chui and A. Rangarajan, "A new point matching algorithm for non-rigid registration," *Computer Vision and Image Understanding*, vol. 89, no. 2-3, pp. 114–141, 2003.

[3] M. Mori, A. Suzuki, A. Shio, and S. Ohtsuka, "Generating new samples from handwriting numerals based on point correspondence," in *Int'l Workshop on Frontiers in Handwriting Recognition*, 2000, pp. 281–290.

[4] R. Plamondon, "A renaissance of handwriting," *Machine Vision and Application*, vol. 8, pp. 195–196, 1995.

[5] G.P. van Galen and G.E. Stelmach, "Handwriting: Issues of psychomotor control and cognitive models," *Special Issue of Acta Psychologica*, vol. 82, no. 1-3, 1993.

[6] R. Plamondon and S. Srihari, "On-line and off-line handwriting recognition: A comprehensive survey," *IEEE Trans. Pattern Anal. Machine Intell.*, vol. 22, no. 1, pp. 63–84, 2000.

[7] J. Wan, A.M. Wing, and N. Sovik, *Development of Graphic Skills: Research, Perspective, and Educational Implications.* Academic Press, London, 1991.

[8] M. Simner, W. Hulstijn, and P. Girouard, "Forensic, developmental and neuropsychological aspects of handwriting," *Special Issue of Journal of Forensic Document Examination*, 1994.

[9] J. Illingworth and J. Kittler, "A survey of the Hough transform," *Computer Vision, Graphics, and Image Processing*, vol. 44, pp. 87–116, 1988.

[10] J. Liu, X. Ding, and Y. Wu, "Description and recognition of form and automated form data entry," in *Proc. Int'l Conf. Document Analysis and Recognition*, 1995, pp. 579–582.

[11] B. Yu and A.K. Jain, "A generic system for form dropout," *IEEE Trans. Pattern Anal. Machine Intell.*, vol. 18, no. 11, pp. 1127–1131, 1996.

[12] D. Dori and W. Liu, "Sparse pixel vectorization: An algorithm and its performance evaluation," *IEEE Trans. Pattern Anal. Machine Intell.*, vol. 21, no. 3, pp. 202–215, 1999.

[13] J.-L. Chen and H.-J. Lee, "An efficient algorithm for form structure extraction using strip projection," *Pattern Recognition*, vol. 31, no. 9, pp. 1353–1368, 1998.

[14] O. Hori and D. Doermann, "Robust table-form structure analysis based on box-driven reasoning," in *Proc. Int'l Conf. Document Analysis and Recognition*, 1995, pp. 218–221.

[15] Y.Y. Tang, C.Y. Suen, and C.D. Yan, "Financial document processing based on staff line and description language," *IEEE Trans. Systems, Man and Cybernetics*, vol. 25, no. 5, pp. 738–753, 1995.

[16] F. Cesarini, M. Gori, and S. Marinai, "INFORMys: A flexible invoice-like form-reader system," *IEEE Trans. Pattern Anal. Machine Intell.*, vol. 20, no. 7, pp. 730–745, 1998.

[17] A.K. Jain and B. Yu, "Document representation and its application to page decomposition," *IEEE Trans. Pattern Anal. Machine Intell.*, vol. 20, no. 3, pp. 294–308, 1998.

[18] L. O'Gorman, "The document spectrum for page layout analysis," *IEEE Trans. Pattern Anal. Machine Intell.*, vol. 15, no. 11, pp. 1162–1173, 1993.

[19] K.C. Fan, L.S. Wang, and Y.T. Tu, "Classification of machine-printed and handwritten texts using character block layout variance," *Pattern Recognition*, vol. 31, no. 9, pp. 1275–1284, 1998.

[20] J. Fanke and M. Oberlander, "Writing style detection by statistical combination of classifier in form reader applications," in *Proc. Int'l Conf. Document Analysis and Recognition*, 1993, pp. 581–585.

[21] V. Pal and B.B. Chaudhuri, "Machine-printed and handwritten text lines identification," *Pattern Recognition Letters*, vol. 22, no. 3-4, pp. 431–441, 2001.

[22] S.N. Srihari, Y.C. Shim, and V. Ramanprasad, "A system to read names and address on tax forms," CEDAR, SUNY, Buffalo, NY, Tech. Rep. CEDAR-TR-94-2, 1994.

[23] J.K. Guo and M.Y. Ma, "Separating handwritten material from machine printed text using hidden Markov models," in *Proc. Int'l Conf. Document Analysis and Recognition*, 2001, pp. 439–443.

[24] K. Kuhnke, L. Simoncini, and Z.M. Kovacs-V, "A system for machine-written and hand-written character distinction," in *Proc. Int'l Conf. Document Analysis and Recognition*, 1995, pp. 811–814.

[25] Y. Zheng, H. Li, and D. Doermann, "The segmentation and identification of handwriting in noisy document images," in *Proc. Int'l Workshop on Document Analysis Systems*, 2002, pp. 95–105.

[26] S. Belongie, J. Malik, and J. Puzicha, "Shape matching and object recognition using shape contexts," *IEEE Trans. Pattern Anal. Machine Intell.*, vol. 24, no. 4, pp. 509–522, 2002.

[27] S. Naoti, M. Suwa, M. Yabuki, and Y. Hotta, "Global interpolation in the segmentation of handwritten characters overlapping a border," *IEICE Trans. Inf. & Syst.*, vol. E78-D, no. 7, pp. 909–916, 1995.

[28] M. Cheriet, J.N. Said, and C.Y. Suen, "A formal modal for document processing of business forms," in *Proc. Int'l Conf. Document Analysis and Recognition*, 1995, pp. 210–213.

[29] D. Guillevic and C.Y. Suen, "Cursive script recognition: A fast reader scheme," in *Proc. Int'l Conf. Document Analysis and Recognition*, 1993, pp. 311–314.

[30] J.Y. Yoo, M. Kim, S.Y. Han, and Y.B. Kwon, "Line removal and restoration of handwritten characters on the form documents," in *Proc. Int'l Conf. Document Analysis and Recognition*, 1997, pp. 128–131.

[31] Y. Chung, K. Lee, J. Yaik, and Y. Lee, "Extraction and restoration of digits touching or overlapping lines," in *Proc. Int'l Conf. Pattern Recognition*, 1996, pp. 155–159.

[32] D. Doermann and A. Rosenfeld, "The processing of form documents," in *Proc. Int'l Conf. Document Analysis and Recognition*, 1993, pp. 497–501.

[33] M.D. Garris, "Intelligent form removal with character stroke preservation," in *Proc. SPIE Conf. Document Recognition*, 1996, pp. 321–332.

[34] Y. Zheng, C. Liu, and X. Ding, "Form frame line removal with line width threshold approach," *Pattern Recognition and Artificial Intelligence (in Chinese)*, vol. 14, no. 2, pp. 210–214, 2001.

[35] Y. Zheng, "A study on generic form document processing system," Master's thesis, Tsinghua University, Beijing, China, July 2001.

[36] D. Dori, Y. Liang, and J. Dowell, "Sparse-pixel recognition of primitives in engineering drawings," *Machine Vision and Application*, vol. 6, pp. 69–82, 1993.

[37] D. Blostein and H.S. Baird, "A critical survey of music image analysis," in *Structured Document Image Analysis*, H.S. Baird, H. Bunke, and K. Yamamoto, Eds. Springer-Verlag, 1992, pp. 405–434.

[38] P.V.C. Hough, "Machine analysis of bubble chamber pictures," in *Int'l Conf. High Energy Accelerators and Instrumentation*, 1959.

[39] H. Tamura, "A comparison of line thinning algorithms from digital geometry viewpoint," in *Proc. Int'l Conf. Pattern Recognition*, 1978, pp. 715–719.

[40] J. Jimenez and J.L. Navalon, "Some experiments in image vectorization," *IBM J. Res. Develop.*, vol. 26, pp. 724–734, 1982.

[41] O. Hori and S. Tanigawa, "Raster-to-vector conversion by line fitting based on contours and skeletons," in *Proc. Int'l Conf. Document Analysis and Recognition*, 1993, pp. 353–358.

[42] A.K. Chhabra, V. Misra, and J. Arias, "Detection of horizontal lines in noisy run length encoded images: The FAST method," in *Proc. IAPR Int'l Workshop on Graphics Recognition*, 1995, pp. 35–48.

[43] J.F. Arias, A. Chhabra, and V. Misra, "Finding straight lines in drawings," in *Proc. Int'l Conf. Document Analysis and Recognition*, 1997, pp. 788–791.

[44] V.P. d'Andecy, J. Camillerapp, and I. Leplumey, "Kalman filtering for segment detection: Application to music score analysis," in *Proc. Int'l Conf. Pattern Recognition*, 1994, pp. 301–305.

[45] J.F. Arias, C.P. Lan, S. Surya, R. Kasturi, and A. Chhabra, "Interpretation of telephone system manhole drawings," *Pattern Recognition Letters*, vol. 16, no. 4, pp. 355–369, 1995.

[46] D. Dori, L. Wenyin, and M. Peleg, "How to win a dashed line detection contest," in *Proc. IAPR Int'l Workshop on Graphics Recognition*, 1997, pp. 286–300.

[47] J.W. Roach and J.E. Tatem, "Using domain knowledge in low-level visual processing to interpret handwritten music: An experiment," *Pattern Recognition*, vol. 21, no. 1, pp. 33–44, 1988.

[48] J.J. Hull, "Document image skew detection: Survey and annotated bibliography," in *Document Analysis Systems II*, J.J. Hull and S.L. Taylor, Eds. Word Scientific, 1998.

[49] S.W. Lam, L. Javanbakht, and S.N. Srihari, "Anatomy of a form reader," in *Proc. Int'l Conf. Document Analysis and Recognition*, 1993, pp. 506–509.

[50] T. Kanungo and R.M. Haralick, "An automatic closed-loop methodology for generating character groundtruth for scanned documents," *IEEE Trans. Pattern Anal. Machine Intell.*, vol. 21, no. 2, pp. 179–183, 1999.

[51] L.R. Rabiner, "A tutorial on hidden Markov models and selected application in speech recognition," *Proc. IEEE*, vol. 77, no. 2, pp. 257–286, 1989.

[52] G. Grimmett and D. Stirzaker, *Probability and Random Processes*, 2nd ed. Oxford University Press, 2001.

[53] S.E. Levinson, "Continuously variable duration hidden Markov models for automatic speech recognition," *Computer, Speech and Language*, vol. 1, no. 1, pp. 29–45, 1986.

[54] D. Rumelhart and J. McClelland, *Parallel Distributed Processing: Explorations in the Microstructure of Cognition.* MIT Press, Cambridge, MA, 1986.

[55] K.J. Lang, A.H. Waibel, and G.E. Hinton, "A time-delay neural network architecture for isolated word recognition," *Neural Networks*, vol. 3, pp. 23–43, 1990.

[56] D. Jurafsky and J. Martin, *Speech and Language Processing: An Introduction to Natural Language Processing, Computational Linguistics, and Speech Recognition.* Prentice Hall, New Jersey, 2000.

[57] J. Nelder and R. Mead, "A simplex method for function minimization," *Computer Journal*, vol. 7, pp. 308–313, 1965.

[58] W. Liu and D. Dori, "A protocol for performance evaluation of line detection algorithms," *Machine Vision and Application*, vol. 9, no. 5-6, pp. 57–68, 1997.

[59] D.P. Huttenlocher, G.A. Klanderman, and W.J. Rucklidge, "Comparing images using the Hausdorff distance," *IEEE Trans. Pattern Anal. Machine Intell.*, vol. 15, no. 9, pp. 850–863, 1993.

[60] B. Kong, I.T. Phillips, and R.M. Haralick, "A benchmark: Performance evaluation of dashed-line detection algorithm," in *Proc. IAPR Int'l Workshop on Graphics Recognition*, 1995, pp. 270–285.

[61] L.Y. Tseng and R.C. Chen, "Recognition and data extraction of form documents based on three types of line segments," *Pattern Recognition*, vol. 31, no. 10, pp. 1525–1540, 1998.

[62] A. Ting and M. Leung, "Form recognition using linear structure," *Pattern Recognition*, vol. 32, no. 4, pp. 645–656, 1999.

[63] J. Liu and A.K. Jain, "Image-based form document retrieval," *Pattern Recognition*, vol. 33, no. 3, pp. 503–513, 2000.

[64] T. Kanungo, R.M. Haralick, and I. Phillips, "Nonlinear local and global document degradation models," *Int'l J. Imaging Systems and Technology*, vol. 5, no. 4, pp. 220–230, 1994.

[65] D.L. Dimmick, M.D. Garris, and C.L. Wilson. NIST structured forms reference set. [Online]. Available: http://www.nist.gov/srd/nistsd2.htm

[66] J.J. Hull, "Incorporating language syntax in visual text recognition with a statistical model," *IEEE Trans. Pattern Anal. Machine Intell.*, vol. 18, no. 12, pp. 1251–1256, 1996.

[67] R.M.K. Sinha, B. Prasada, G.F. Houles, and M. Sabourin, "Hybrid contextual text recognition with string matching," *IEEE Trans. Pattern Anal. Machine Intell.*, vol. 15, no. 9, pp. 915–925, 1993.

[68] R. Chellappa and A. Jain, Eds., *Markov Random Fields: Theory and Application*. Academic Press, San Diego, 1993.

[69] S. Geman and D. Geman, "Stochastic relaxation, Gibbs distribution and the Bayesian restoration of images," *IEEE Trans. Pattern Anal. Machine Intell.*, vol. 6, no. 6, pp. 721–741, 1984.

[70] S.Z. Li, *Markov Random Field Modeling in Image Analysis*, 2nd ed. Springer-Verlag, New York, 2001.

[71] Y. Zheng, C. Liu, and X. Ding, "Single character type identification," in *Proc. SPIE Conf. Document Recognition and Retrieval*, 2002, pp. 49–56.

[72] H.S. Baird, "Calibration of document image defect models," in *Proc. Ann. Symp. Document Analysis and Information Retrieval*, 1993, pp. 1–16.

[73] S. Sural and P.K. Das, "A two-state Markov chain model of degraded document images," in *Proc. Int'l Conf. Document Analysis and Recognition*, 1999, pp. 463–466.

[74] M. Cannon, J. Hochberg, and P. Kelly, "Quality assessment and restoration of typewritten document images," *Int'l J. Document Analysis and Recognition*, vol. 2, pp. 80–89, 1999.

[75] H. Li and D. Doermann, "Text quality estimation in video," in *Proc. SPIE Conf. Document Recognition and Retrieval*, 2002, pp. 232–243.

[76] J. Liang, I.T. Phillips, and R.M. Haralick, "Performance evaluation of document layout analysis algorithms on the UW data set," in *Proc. SPIE Conf. Document Recognition*, 1997, pp. 149–160.

[77] R.P. Loce and E.R. Dougherty, *Enhancement and Restoration of Digital Documents – Statistical Design of Nonlinear Algorithms*. SPIE Optical Engineering Press, 1997.

[78] L. O'Gorman, "Image and document processing techniques for the RightPages electronic library system," in *Proc. Int'l Conf. Pattern Recognition*, 1992, pp. 820–825.

[79] K. Chinnasarn, Y. Rangsanseri, and P. Thitimajshima, "Removing salt-and-pepper noise in text/graphics images," in *Proc. IEEE Asia-Pacific Conf. Circuits and Systems*, 1998, pp. 459–462.

[80] J. Liang and R.M. Haralick, "Document image restoration using binary morphological filters," in *Proc. SPIE Conf. Document Recognition*, 1996, pp. 274–285.

[81] T. Kanungo, R.M. Haralick, H.S. Baird, W. Stuetzle, and D. Madigan, "Document degradation models: Parameter estimation and model validation," in *Proc. Int'l Workshop on Machine Vision Applications*, 1994, pp. 552–557.

[82] T. Kanungo, H.S. Baird, and R.M. Haralick, "Validation and estimation of document degradation models," in *Proc. Ann. Symp. Document Analysis and Information Retrieval*, 1995, pp. 217–228.

[83] T. Kanungo and Q. Zheng, "Estimation of morphological degradation model parameters," in *Proc. IEEE Int'l Conf. Speech and Signal Processing*, 2001, pp. 1961–1964.

[84] H.S. Baird, "Document image quality: Making fine discriminations," in *Proc. Int'l Conf. Document Analysis and Recognition*, 1999, pp. 459–462.

[85] K.-C. Fan, Y.-K. Wang, and T.-R. Lay, "Marginal noise removal of document images," *Pattern Recognition*, vol. 35, no. 11, pp. 2593–2611, 2002.

[86] H. Nishida and T. Suzuki, "Correcting show-through effects on document images by multiscale analysis," in *Proc. Int'l Conf. Pattern Recognition*, 2002.

[87] C.L. Tan, R. Lao, and P. Shen, "Restoration of archival document using a wavelet technique," *IEEE Trans. Pattern Anal. Machine Intell.*, vol. 24, no. 10, pp. 1399–1404, 2002.

[88] D. Gabor, "Theory of communication," *J. Inst. Elect. Engr.*, vol. 93, pp. 429–459, 1946.

[89] A.K. Jain and S. Bhattacharjee, "Text segmentation using Gabor filters for automatic document processing," *Machine Vision and Application*, vol. 5, pp. 169–184, 1992.

[90] T. Akiyama and N. Hagita, "Automated entry system for printed documents," *Pattern Recognition*, vol. 23, no. 11, pp. 1141–1154, 1990.

[91] R. Haralick, B. Shanmugam, and I. Dinstein, "Texture features for image classification," *IEEE Trans. System, Man and Cybernetics*, vol. 3, no. 6, pp. 610–622, 1973.

[92] A. Soffer, "Image categorization using texture features," in *Proc. Int'l Conf. Document Analysis and Recognition*, 1997, pp. 233–237.

[93] K. Fukunaga, *Introduction to Statistical Pattern Recognition*, 2nd ed. Academic Press, New York, 1990.

[94] A.K. Jain and D. Zongker, "Feature selection: Evaluation, application and small sample performance," *IEEE Trans. Pattern Anal. Machine Intell.*, vol. 19, no. 2, pp. 153–158, 1997.

[95] L. Kukolick and R. Lippmann. LNKnet user's guide. [Online]. Available: http://www.ll.mit.edu/IST/lnknet/

[96] X. Lin, X. Ding, and M. Chen, "Adaptive confidence transform based classifier combination for Chinese character recognition," *Pattern Recognition Letters*, vol. 19, no. 10, pp. 975–988, 1998.

[97] C. Wolf and D. Doermann, "Binarization of low quality text using a Markov random field model," in *Proc. Int'l Conf. Pattern Recognition*, 2002.

[98] P.B. Chou, P.R. Cooper, M.J. Swain, C.M. Brown, and L.E. Wixson, "Probabilistic network inference for cooperative high and low level vision," in *Markov Random Fields: Theory and Application*, R. Chellappa and A.K. Jain, Eds. Academic Press, San Diego, 1993, pp. 211–243.

[99] V. Vapnik, *The Nature of Statistical Learning Theory.* Springer-Verlag, New York, 1995.

[100] C.-C. Chang and C.-J. Lin. Libsvm – A library for support vector machines. [Online]. Available: http://www.csie.ntu.edu.tw/~cjlin/libsvm/

[101] ScanSoft Corp. ScanSoft developer's kit 2000. [Online]. Available: http://www.scansoft.com

[102] J. Kanai, S.V. Rice, T.A. Nartker, and G. Nagy, "Automated evaluation of OCR zoning," *IEEE Trans. Pattern Anal. Machine Intell.*, vol. 17, no. 1, pp. 86–90, 1995.

[103] S. Mao and T. Kanungo, "Empirical performance evaluation methodology and its application to page segmentation algorithms," *IEEE Trans. Pattern Anal. Machine Intell.*, vol. 23, no. 3, pp. 242–256, 2001.

[104] S. Randriamasy, L. Vincent, and B. Wittner, "An automatic benchmarking scheme for page segmentation," in *Proc. SPIE Conf. Document Recognition*, 1994, pp. 217–227.

[105] T. Wakahara, "Shape matching using LAT and its application to handwritten numeral recognition," *IEEE Trans. Pattern Anal. Machine Intell.*, vol. 16, no. 6, pp. 618–629, 1994.

[106] T. Wakahara and K. Odaka, "Adaptive normalization of handwritten characters using global/local affine transform," *IEEE Trans. Pattern Anal. Machine Intell.*, vol. 20, no. 12, pp. 1332–1341, 1998.

[107] Y. Zheng and D. Doermann, "Handwriting matching and its application to handwriting synthesis," in *Proc. Int'l Conf. Document Analysis and Recognition*, 2005, pp. 861–865.

[108] T. Varga and H. Bunke, "Generation of synthetic training data for an HMM-based handwriting recognition system," in *Proc. Int'l Conf. Document Analysis and Recognition*, 2003, pp. 618–622.

[109] S. Loncaric, "A survey of shape analysis techniques," *Pattern Recognition*, vol. 31, no. 8, pp. 983–1001, 1998.

[110] R.C. Velkamp and M. Hagedoorn, "State of the art in shape matching," Utrecht University, Netherlands, Tech. Rep. UU-CS-1999-27, 1999.

[111] P.J. Besl and N.D. McKay, "A method for registration of 3-D shapes," *IEEE Trans. Pattern Anal. Machine Intell.*, vol. 14, no. 2, pp. 239–256, 1992.

[112] H. Holstein and B. Li, "Low density feature point matching for articulated pose identification," in *Proc. British Machine Vision Conference*, 2002, pp. 678–687.

[113] P. David, D. DeMenthon, R. Duraiswami, and H.J. Samet, "SoftPOSIT: Simultaneous pose and correspondence determination," *Int. J. Computer Vision*, vol. 59, no. 3, pp. 259–284, 2004.

[114] A. Rangarajan, H. Chui, E. Mjolsness, S. Pappu, L. Davachi, P.S. Goldman-Rakic, and J.S. Duncan, "A robust point matching algorithm for autoradiograph alignment," *Medical Image Analysis*, vol. 4, no. 1, pp. 379–398, 1997.

[115] Z. Zhang, "Iterative point matching for registration of free-form curves and surfaces," *Int. J. Computer Vision*, vol. 13, no. 2, pp. 119–152, 1994.

[116] J. Feldmar and N. Anyche, "Rigid, affine and locally affine registration of free-form surfaces," *Int. J. Computer Vision*, vol. 18, no. 2, pp. 99–119, 1996.

[117] T.B. Sebastian, P.N. Klein, and B.B. Kimia, "On aligning curves," *IEEE Trans. Pattern Anal. Machine Intell.*, vol. 25, no. 1, pp. 116–124, 2003.

[118] I. Biederman, "Recognition by components: A theory of human image understanding," *Psychological Review*, vol. 94, no. 2, pp. 115–147, 1987.

[119] D.J. Field, A. Hayes, and R.E. Hess, "Contour integration by the human visual system: Evidence for a local 'association field'," *Vision Research*, vol. 33, no. 2, pp. 173–193, 1993.

[120] S. Gold and A. Rangarajan, "A graduated assignment algorithm for graph matching," *IEEE Trans. Pattern Anal. Machine Intell.*, vol. 18, no. 4, pp. 377–388, 1996.

[121] A. Rosenfeld, R.A. Hummel, and S.W. Zucker, "Scene labeling by relaxation operations," *IEEE Trans. System, Man and Cybernetics*, vol. 6, no. 6, pp. 420–433, 1976.

[122] L.G. Brown, "A survey of image registration techniques," *ACM Computing Survey*, vol. 24, no. 4, pp. 325–376, 1992.

[123] B. Li, Q. Meng, and H. Holstein, "Point pattern matching and applications – a review," in *Proc. Int'l Conf. Systems, Man and Cybernetics*, 2003, pp. 729–736.

[124] F.L. Bookstein, "Principal warps: Thin-plate splines and the decomposition of deformation," *IEEE Trans. Pattern Anal. Machine Intell.*, vol. 11, no. 6, pp. 567–585, 1989.

[125] A.L. Yuille and N.M. Grzywacz, "A mathematical analysis of the motion coherence theory," *Int. J. Computer Vision*, vol. 3, no. 2, pp. 155–175, 1989.

[126] R. Jonker and A. Volgenant, "A shortest augmenting path algorithm for dense and sparse linear assignment problems," *Computing*, vol. 38, no. 4, pp. 325–340, 1987.

[127] J. Glaunes, A. Trouvé, and L. Younes, "Diffeomorphic matching of distributions: A new approach for unlabelled point-sets and sub-manifolds matching," in *Proc. IEEE Conf. Computer Vision and Pattern Recognition*, 2004, pp. 712–718.

[128] W.J. Christmas, J. Kittler, and M. Petrou, "Structural matching in computer vision using probabilistic relaxation," *IEEE Trans. Pattern Anal. Machine Intell.*, vol. 17, no. 8, pp. 749–764, 1995.

[129] R.C. Wilson and E.R. Hancock, "Structural matching by discrete relaxation," *IEEE Trans. Pattern Anal. Machine Intell.*, vol. 19, no. 6, pp. 634–648, 1997.

[130] D. Conte, P. Foggia, C. Sansone, and M. Vento, "Thirty years of graph matching in pattern recognition," *Int'l J. Pattern Recognition and Artificial Intelligence*, vol. 18, no. 3, pp. 265–298, 2004.

[131] C.S. Kenney, B.S. Manjunath, M. Zuliani, G.A. Hewer, and A.V. Nevel, "A condition number for point matching with application to registration and postregistration error estimation," *IEEE Trans. Pattern Anal. Machine Intell.*, vol. 25, no. 11, pp. 1437–1454, 2003.

[132] J. Kittler and J. Illingworth, "Relaxation labeling algorithms – a review," *Image and Vision Computing*, vol. 3, no. 4, pp. 206–216, 1985.

[133] J. Kittler and E.R. Hancock, "Combining evidence in probabilistic relaxation," *Int'l J. Pattern Recognition and Artificial Intelligence*, vol. 3, no. 1, pp. 29–51, 1989.

[134] O.D. Faugeras and M. Berthod, "Improving consistency and reducing ambiguity in stochastic labeling: An optimization approach," *IEEE Trans. Pattern Anal. Machine Intell.*, vol. 3, no. 4, pp. 412–424, 1981.

[135] R.A. Hummel and S.W. Zucker, "On the foundations of relaxation labeling processes," *IEEE Trans. Pattern Anal. Machine Intell.*, vol. 5, no. 3, pp. 267–287, 1983.

[136] M. Pelillo, "The dynamics of nonlinear relaxation labeling processes," *J. Mathematical Imaging and Vision*, vol. 7, no. 4, pp. 309–323, 1997.

[137] K.E. Price, "Relaxation matching techniques – a comparison," *IEEE Trans. Pattern Anal. Machine Intell.*, vol. 7, no. 5, pp. 617–623, 1985.

[138] L.S. Davis, "Shape matching using relaxation techniques," *IEEE Trans. Pattern Anal. Machine Intell.*, vol. 1, no. 1, pp. 60–72, 1979.

[139] O.D. Faugeras and K.E. Price, "Semantic description of aerial images using stochastic labeling," *IEEE Trans. Pattern Anal. Machine Intell.*, vol. 3, no. 6, pp. 633–642, 1981.

[140] S.Z. Li, "Matching: Invariant to translations, rotations, and scale changes," *Pattern Recognition*, vol. 25, no. 6, pp. 583–594, 1992.

[141] S. Ranade and A. Rosenfeld, "Point pattern matching by relaxation," *Pattern Recognition*, vol. 12, no. 4, pp. 269–275, 1980.

[142] J. Ton and A.K. Jain, "Registering Landsat images by point matching," *IEEE Trans. Geoscience and Remote Sensing*, vol. 27, no. 5, pp. 642–651, 1989.

[143] R. Sinkhorn, "A relationship between arbitrary positive matrices and doubly stochastic matrices," *The Annals of Mathematical Statistics*, vol. 35, no. 2, pp. 876–879, 1964.

[144] M. Pelillo, "Replicator equations, maximal cliques, and graph isomorphism," *Neural Computation*, vol. 11, no. 8, pp. 1933–1955, 1999.

[145] I.L. Dryden and K.V. Mardia, *Statistical Shape Analysis*. John Wiley, Chichester, 1998.

[146] P.J. Rousseeuw and A.M. Leroy, *Robust Regression and Outlier Detection*. John Wiley and Sons, New York, 1987.

[147] D.L. Donoho and P.J. Huber, "The notion of breakdown point," in *A Festschrift for Erich L. Lehmann Belmont*, P.J. Bickel, K.A. Doksum, and J.L. Hodges, Eds. Wadsworth, CA, 1983, pp. 157–184.

[148] N. Duta, A.K. Jain, and K.V. Mardia, "Matching of palmprint," *Pattern Recognition Letters*, vol. 23, no. 4, pp. 477–485, 2002.

[149] G. Wahba, *Spline Models for Observational Data*. Soc. Industrial and Applied Math., 1990.

[150] F. Girosi, M. Jones, and T. Poggio, "Regularization theory and neural networks architectures," *Neural Computation*, vol. 7, no. 2, pp. 219–269, 1995.

[151] G.K. Kanji, *100 Statistical Tests*. Sage Publications, 1999.

[152] J. Wang, C. Wu, Y.-Q. Xu, and H.-Y. Shum, "Combining shape and physical models for online cursive handwriting synthesis," *Int'l J. Document Analysis and Recognition*, Online First, 2004.

[153] H. Choi, S.-J. Cho, and J.H. Kim, "Generation of handwritten characters with Bayesian network based on-line handwriting recognizers," in *Proc. Int'l Conf. Document Analysis and Recognition*, 2003, pp. 995–999.

[154] T.M. Rath and R. Manmatha, "Word image matching using dynamic time warping," in *Proc. IEEE Conf. Computer Vision and Pattern Recognition*, 2003, pp. 521–527.

[155] S. Kuo and O.E. Agazzi, "Keyword spotting in poorly printed documents using pseudo 2-d hidden Markov models," *IEEE Trans. Pattern Anal. Machine Intell.*, vol. 16, no. 8, pp. 842–848, 1994.

[156] F.R. Chen, D.S. Bloomberg, and L.D. Wilcox, "Detection and location of multi-character sequences in lines of imaged text," *J. Electronic Imaging*, vol. 5, no. 1, pp. 37–49, 1996.

[157] J.L. Rothfeder, S. Feng, and T.M. Rath, "Using corner feature correpondences to rank work images by similarity," in *IEEE Workshop on Document Image Analysis and Retrieval*, 2003.

[158] H. Ling and D.W. Jacobs, "Using the inner-distance for classification of articu-lated shapes," in *Proc. IEEE Conf. Computer Vision and Pattern Recognition*, 2005, pp. 719–726.

[159] K. Siddiqi, A. Shokoufandeh, S.J. Dickinson, and S.W. Zucker, "Shock graphs and shape matching," *Int. J. Computer Vision*, vol. 35, no. 1, pp. 13–32, 1999.

[160] T.B. Sebastian, P.N. Klein, and B. Kimia, "Recognition of shapes by editing their shock graphs," *IEEE Trans. Pattern Anal. Machine Intell.*, vol. 26, no. 5, pp. 550–571, 2004.

[161] L.J. Latecki, R. Lakamper, and U. Eckhardt, "Shape descriptors for non-rigid shapes with a single closed contour," in *Proc. IEEE Conf. Computer Vision and Pattern Recognition*, 2000, pp. 424–429.

[162] E.G.M. Petrakis, A. Diplaros, and E. Milios, "Matching and retrieval of distorted and occluded shapes using dynamic programming," *IEEE Trans. Pattern Anal. Machine Intell.*, vol. 24, no. 11, pp. 1501–1516, 2002.

[163] D.S. Doermann and A. Rosenfeld, "Recovery of temporal information from static images of handwriting," *Int. J. Computer Vision*, vol. 52, no. 1-2, pp. 143–164, 1994.

[164] D.G. Kendall, "The diffusion of shape," *Advances in Applied Probability*, vol. 9, pp. 428–430, 1977.

[165] C.R. Dyer and A. Rosenfeld, "Thinning algorithms for gray-scale pictures," *IEEE Trans. Pattern Anal. Machine Intell.*, vol. 1, no. 1, pp. 88–89, 1979.

[166] T.Y. Zhang and C.Y. Suen, "A fast parallel algorithm for thinning digital patterns," *Communications of the ACM*, vol. 27, no. 3, pp. 236–239, 1984.

Wissenschaftlicher Buchverlag bietet

kostenfreie

Publikation

von

wissenschaftlichen Arbeiten

Diplomarbeiten, Magisterarbeiten, Master und Bachelor Theses
sowie Dissertationen, Habilitationen und wissenschaftliche Monographien

Sie verfügen über eine wissenschaftliche Abschlußarbeit zu aktuellen oder zeitlosen
Fragestellungen, die hohen inhaltlichen und formalen Ansprüchen genügt,
und haben **Interesse an einer honorarvergüteten Publikation**?

Dann senden Sie bitte erste Informationen über Ihre Arbeit per Email
an info@vdm-verlag.de. Unser Außenlektorat meldet sich umgehend bei Ihnen.

VDM Verlag Dr. Müller Aktiengesellschaft & Co. KG
Dudweiler Landstraße 125a
D - 66123 Saarbrücken

www.vdm-verlag.de